GUIDE to RESEARCH PROJECTS FOR ENGINEERING STUDENTS

Planning, Writing and Presenting

Eng-Choon Leong

Nanyang Technological University, Singapore

Carmel Lee-Hsia Heah

Nanyang Technological University, Singapore

Kenneth Keng Wee Ong

Nanyang Technological University, Singapore

CRC Press
Taylor & Francis Group
Boca Raton London New York

CRC Press is an imprint of the
Taylor & Francis Group, an **Informa** business

A SPON PRESS BOOK

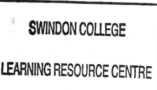

SWINDON COLLEGE

LEARNING RESOURCE CENTRE

CRC Press
Taylor & Francis Group
6000 Broken Sound Parkway NW, Suite 300
Boca Raton, FL 33487-2742

© 2016 by Taylor & Francis Group, LLC
CRC Press is an imprint of Taylor & Francis Group, an Informa business

No claim to original U.S. Government works

International Standard Book Number-13: 978-1-4822-3877-8 (Paperback)

Library of Congress Cataloging-in-Publication Data

Leong, E. C.
   Guide to research projects for engineering students : planning, writing and presenting / authors, Eng-Choon Leong, Carmel Lee-Hsia Heah, Kenneth Keng Wee Ong.
      pages cm
   Includes bibliographical references and index.
   ISBN 978-1-4822-3877-8 (alk. paper)
   1. Engineering--Research. I. Heah, Carmel Lee Hsia II. Ong, Kenneth Keng Wee. III. Title.

TA160.L46 2015
620.0072--dc23   2015007216

Visit the Taylor & Francis Web site at
http://www.taylorandfrancis.com

and the CRC Press Web site at
http://www.crcpress.com

GUIDE to RESEARCH PROJECTS FOR ENGINEERING STUDENTS

Planning, Writing and Presenting

Contents

15 Writing the results and discussion   115

16 Writing the conclusion   127

17 Writing the abstract and front matter   135

List of figures

List of tables

List of writing guidelines

Preface

Why this book? With the Internet, information on technical writing and on conducting research is widely available. However, for students embarking on their *first* research project, sieving through the vast amount of information on the Internet can be a daunting and time-consuming task. There are indeed many books which have been published focusing on either technical writing or on research methodology. Project work, however, requires not just an understanding of research methodology, but also requires complementary skills in information search, technical writing, oral presentation as well as software application skills. Hence, this book adopts an integrated approach which combines research skills and the complementary skills mentioned. This book has been designed with the specific needs of engineering and technical sciences students in mind by addressing their need for clear and practical guidelines on conducting and presenting their research projects.

The book is organised into three parts: Part I (*Planning*), Part II (*Writing*) and Part III (*Presenting*). Each part is broken down into a number of concise chapters that focus on a specific set of skills that students will need in conducting their research. Part I (*Planning*) includes topics on choosing a research project, researching a topic, writing a research project proposal, planning a research project, research methodology and research methods, and keeping research records. Part II (*Writing*) covers topics on starting to write, grammar, punctuation and word usage, the do's and don'ts of technical writing, strategies for writing a good report or thesis, writing the introduction and literature review, writing methods and materials, writing results and discussions, writing the conclusion, writing the abstract and front matter, referencing, using sources and avoiding plagiarism. Part III (*Presenting*) covers topics on how to create figures and layout, preparing for an oral presentation, and the do's and don'ts of oral presentations.

The use of information technology (IT) tools is indispensable in a research project. For most students, the personal computer is an indispensable tool for conducting a research project and in writing and presenting their research. In this book, IT skills in writing and presentation, which are rarely discussed in technical writing books, are integrated into the appropriate chapters.

Chapter 2 (*Frequently Asked Questions [FAQs]*) precedes the three main parts. This chapter has been designed for students who are looking for answers to specific questions. Directions are given to chapters or parts in the book which provide the answers. This referencing method will also be helpful to students who are looking for assistance with a specific task.

The goals of this book are to be immensely practical and to be a self-contained guide. For students who wish to explore specific skills more deeply, references are provided in each chapter. This book has an accompanying website, www.ResearchProjectsforEngineeringStudents.wordpress .com, where readers are welcome to give their feedback.

Acknowledgments

We owe our inspiration for writing this book to our many students, past and present, with whom we have worked with on research projects. By undertaking their respective projects and postgraduate theses, these students have contributed greatly to our understanding of the difficulties that students encounter in undertaking research work and communicating about it.

Last but not least, we thank our families for their support and understanding during the writing of this book.

Chapter 1

Introduction

This book has been written specially for engineering and technical sciences students, undergraduates as well as postgraduates*, who are required to complete a research project and present the final report or thesis[†] as a partial or sole requirement in a degree programme. Its purpose is to provide a framework for first conducting and then writing clearly and comprehensively about it. In this chapter, we provide an overview of these skills that will be addressed more fully in subsequent chapters.

COGNITIVE SKILLS

Students embarking on a research project are expected to have gained sufficient knowledge in their field of study. They are also expected to be able to work independently on a research project related to their field, that is, to plan and manage their time, conduct the research, write and present their findings. Cognitive skills required for research are very broad-based and encompass skills ranging from searching for information to a specialised analysis of a particular research problem. The cognitive skills involved in an undergraduate or postgraduate research project are at much higher levels than those required for taking courses. The different levels of cognitive skills required for courses and research projects can be shown using Bloom's revised taxonomy of learning domains (Anderson et al., 2000) in Table 1.1. The cognitive skills required for research projects are addressed in the three parts of this book: *Planning*, *Writing* and *Presenting*.

---

* Postgraduate/graduate: Both terms refer to someone who already has an undergraduate degree and is pursuing a higher degree. The former term is used in the United Kingdom and the latter term is used in the United States.

† Thesis/dissertation: Both terms refer to the final written document submitted for a master's or PhD degree. The terms are used interchangeably. In the United Kingdom, 'thesis' is usually associated with a PhD degree whereas 'dissertation' is usually associated with a master's degree. The reverse is true in the United States.

Table 1.1 Bloom's revised taxonomy of learning domains

Definitions	Bloom's definitions	Basic courses	Intermediate courses	Advanced courses	Undergraduate project	Postgraduate project
Knowledge	Remember previously learned information					
Comprehension	Demonstrate an understanding of the facts					
Application	Apply knowledge to actual situations					
Analysis	Break down objects or ideas into simpler parts and find evidence to support generalisations					
Synthesis	Compile component ideas into a new whole or propose alternative solutions					
Evaluation	Make and defend judgments based on internal evidence or external criteria					

ROLE OF SUPERVISOR

While carrying out the research project, the student will be working under a supervisor‡ whose role is to advise the student on the technical content of the research project and to assess the performance of the student. The role of the supervisor is more of a mentor and less of a teacher. The supervisor does not teach but guides the student in the research project. Therefore, it is important for you as a supervisee to discuss your ideas and progress with your supervisor on a regular basis. The supervisor will then be able to help you to refine your ideas as well as suggest solutions when you encounter problems.

WRITING SKILLS

To complete a research project successfully, students also need good language and technical writing skills. We have seen students who have worked diligently on a research project and obtained impressive results but who

---

‡ Supervisor/advisor: Both terms refer to a faculty member who guides a student in a research project. 'Supervisor' is more commonly used in the United Kingdom whereas 'advisor' is more commonly used in the United States.

were let down by poor writing and presentation at the end. Although writing and communication skills courses are part of the engineering curriculum in most universities, many students have difficulty in transferring what they have learnt from these courses to an engineering research project, especially if these courses are taken much earlier than when these skills are needed. The chapters on writing aim to address this need by adopting a highly practical approach as writing skills, and language conventions are introduced and explained in the context of the chapters on the report or thesis in which these skills and conventions apply.

INFORMATION AND TECHNOLOGY SKILLS

Finally, research projects require the use of a certain degree of office productivity software. At the most basic level, a research project will require an office suite consisting of word-processing software, charting software and presentation software. Several options are available, for example Microsoft Office, Open Office, Apache Open Office, WordPerfect Office, LibreOffice, NeoOffice, ThinkFree Office, Kingsoft Office and so on. There are many books on using specific office productivity software but these books become outdated fairly quickly as new software and new versions of existing software appear in the market. However, there are some features that are common in all office suites. In chapters where information and technology (IT) skills are discussed, we have focused on Microsoft Office as it is the most commonly used suite. For illustration of the software, we will be applying the Pareto principle which states that about 80% of the effects come from 20% of the causes. In other words, we believe that by pointing out to you features of the software, which though useful are not well known, will allow you to become more productive and, hence, you can then devote more time to the technical content of your research project.

In summary, this book has been designed to be a highly practical guide for you when you embark on a research project. Topics covering the full range of skills required in a research project are presented in concise chapters. You can read the book from beginning to end, or specific chapters or sections by using the 'frequently asked questions' in Chapter 2.

We welcome your comments on using this book at www.Research ProjectsforEngineeringStudents.wordpress.com.

REFERENCE

Anderson, L. W., Krathwohl, D. R., Airasian, P. W., Cruikshank, K. A., Mayer, R. E., Pintrich, P. R., Raths, J., & Wittrock, M. C. (2000). *A Taxonomy for Learning, Teaching, and Assessing: A Revision of Bloom's Taxonomy of Educational Objectives*. New York: Pearson, Allyn & Bacon.

Chapter 2

Frequently asked questions (FAQs)

Below are questions frequently asked by students about their research projects. The answers are cross-referenced to the chapters or sections where answers to the questions can be found. Questions commonly fall into one of the following categories: planning, writing, written presentation or oral presentation.

	Chapter (section) where questions are answered
Questions related to planning	
1. What research skills are required for a final year project/postgraduate project?	Chapter 1
2. What is the difference between a thesis and a dissertation?	Chapter 1 (Footnote[†] opening paragraph)
3. How does my supervisor evaluate me?	Chapter 1 (Role of Supervisor)
4. What kind of help can I expect from my supervisor?	Chapter 1 (Role of Supervisor), Chapter 3 (Advice for Undergraduate Students, Advice for Postgraduate Students)
5. What does my supervisor expect of me?	Chapter 1 (Role of Supervisor), Chapter 3 (Advice for Undergraduate Students, Advice for Postgraduate Students)
6. My experiments are not working and time is running out. What can I do?	Chapter 1 (Role of Supervisor), Chapter 6 (Focused on Your Objectives)
7. I find it difficult to get along with my project supervisor. I am afraid he may fail me. What should I do?	Chapter 1 (Role of Supervisor), Chapter 6 (Keep Interested)
8. I have discovered that I have conducted my experiments/research wrongly. What can I do?	Chapter 1 (Role of Supervisor), Chapter 8 (Purpose of Keeping a Research Log)
9. How should I prepare to work on a final year project/postgraduate project?	Chapter 1, Chapter 6 (Project Management)

(Continued)

5

	Chapter (section) where questions are answered
10. How do I go about selecting a final year project/postgraduate project?	Chapter 3
11. Why is it important to meet my project supervisor regularly?	Chapter 3 (Advice for Undergraduate Students, Advice for Postgraduate Students)
12. How do I deal with disagreements with my supervisor?	Chapter 3 (Advice for Undergraduate Students, Advice for Postgraduate Students)
13. How do I carry out research on a topic?	Chapter 4
14. What is a research proposal?	Chapter 5 (Purpose of a Research Proposal)
15. How much time should I allocate to my final year project/postgraduate project?	Chapter 6 (Project Management)
16. What is the difference between methodology and methods?	Chapter 7 (Introduction paragraphs)
17. What is Experimental Research/ Quasi-Experimental Research/Survey Research?	Chapter 7 (Research Methods)
18. How do I keep track of my research?	Chapter 8
19. I cannot get the results that my supervisor expects. What should I do?	Chapter 8 (Purpose of Keeping a Research Log)
20. My project supervisor does not seem to know what I am doing and cannot help me when I have a problem. What should I do?	Chapter 8 (Purpose of Keeping a Research Log), Chapter 6 (Keep Interested), Chapter 3 (Advice for Undergraduate Students, Advice for Postgraduate Students)
21. I have difficulty in finding time to work on my project. What can I do?	Chapter 6
22. How much time should I allocate to writing my report or thesis?	Chapter 9 (Introduction paragraph), Chapter 12
23. When do I start writing my report/ thesis?	Chapter 9 (Starting to Write), Chapter 12 (Start Early)
24. How do I know if I have done enough work for a research project?	Chapter 16 (Introduction paragraph and Checklist for Writing the Conclusion)

Questions related to writing

25. How do I write a research proposal?	Chapter 5
26. I am very poor in grammar. What areas of grammar are important in writing the report or thesis?	Chapter 10 (Common Grammar Problems, Common Sentence Problems)
27. I am not sure how to use punctuation marks such as commas, colons, semicolons, apostrophes and so on. How and when should these punctuation marks be used?	Chapter 10 (Common Punctuation Problems)

(Continued)

	Chapter (section) where questions are answered
28. How can I avoid using the wrong words in my writing?	Chapter 10 (Commonly Misused Words and Phrases)
29. Many words in English have similar spelling. How can I avoid choosing the wrong word?	Chapter 10 (Commonly Misused Words and Phrases)
30. I like research but I am poor in writing. How can I improve my writing skills?	Chapter 10, Chapter 11
31. How do I avoid making grammatical errors in writing my report or thesis?	Chapter 10, Chapter 11
32. What tenses should I use in writing my report or thesis?	Chapter 10, Chapter 13 (Some Key Language Features of an Introduction and Literature Review), Chapter 14 (Some Key Language Features in Methods and Materials), Chapter 15 (Some Key Language Features of Results and Discussion), Chapter 16 (Language Features of the Conclusion), Chapter 17 (Some Key Language Features in an Abstract),
33. I cannot find the motivation to start writing. What should I do?	Chapter 9 (Avoiding Procrastination)
34. What are the commonly confused words in research writing?	Chapter 10 (Pair of Words Commonly Confused)
35. What is formal and informal language?	Chapter 11 (Be Correct)
36. What kind of style should I use to write my report or thesis?	Chapter 11
37. What is a well-written report or thesis?	Chapter 12 (Introduction Paragraph)
38. How many times should I revise my report or thesis?	Chapter 12 (Introduction Paragraph, Start Early, Write the First Draft Quickly)
39. Where can I find guidelines for writing a report or thesis in my institution?	Chapter 12 (Learn from Others), Chapter 22 (Layout of a Report or Thesis, Follow the Guidelines)
40. How do I create a list of tables/figures?	Chapter 12 (Use a Template)
41. When is the best time to write the abstract?	Chapter 12 (Write the First Draft Quickly)
42. How do I write a critical literature review?	Chapter 13
43. How do I write an introduction for my report or thesis?	Chapter 13
44. Should the introduction include citations of published literature?	Chapter 13
45. What should I include in the literature review?	Chapter 13
46. What is a hypothesis statement?	Chapter 13 (Components of an Introduction)

(Continued)

	Chapter (section) where questions are answered
47. What is the difference between an abstract and an introduction?	Chapter 13 (Components of an Introduction), Chapter 17 (Components of an Abstract)
48. How do I write and organise a literature review?	Chapter 13 (Writing the Literature Review as a Separate Section or Chapter, Checklist for Introduction and Literature Review)
49. What is a research hypothesis?	Chapter 13 (Components of an Introduction)
50. What should I write in the materials and methods chapter in my report or thesis?	Chapter 14
51. Do I need to present all the results from my data which I have analysed?	Chapter 15 (Components of Results and Discussion)
52. Could I introduce new literature in the discussion of results?	Chapter 15 (Components of Results and Discussion)
53. How do I write the conclusion for my report/thesis?	Chapter 16
54. Can a conclusion include any new information?	Chapter 16 (Table 16.4)
55. How do I write an abstract for my report or thesis?	Chapter 17
56. Should all the results be presented in the abstract?	Chapter 17
57. What is an informative abstract?	Chapter 17 (Components of an Abstract)
58. What is a descriptive abstract?	Chapter 17 (Components of an Abstract)
59. What is the difference between an informative abstract and a descriptive abstract?	Chapter 17 (Components of an Abstract)
60. What is the difference between an abstract and a summary?	Chapter 17 (Introduction paragraph)
61. What is the typical word count of an abstract?	Chapter 17 (Introduction paragraph)
62. What is 'front matter' in a report or thesis?	Chapter 17 (Parts of Front Matter)
63. I would like to include a figure from a journal paper in my report/thesis. How do I do this?	Chapter 18
64. How do I write a citation?	Chapter 18
65. What referencing system should I use?	Chapter 18 (Features of Two Main Referencing Systems)
66. How can I write so that it is clear which are my ideas and which are other researchers' ideas?	Chapter 18 (In-text Citations), Chapter 19 (Avoiding Plagiarism)

(Continued)

	Chapter (section) where questions are answered
67. I would like to include materials created by others in my report/thesis. How do I do this correctly?	Chapter 18, Chapter 19
68. What is plagiarism?	Chapter 19 (Avoiding Plagiarism)
69. I am afraid I may inadvertently plagiarise published works. How can I avoid this?	Chapter 19 (Avoiding Plagiarism)
70. What is paraphrasing and how do I do it correctly?	Chapter 19 (Avoiding Plagiarism)
71. How do I know if a source is reliable?	Chapter 19 (Evaluating Sources)
72. I find that I am repeating the same thing over and over in my writing. How can I avoid repetition?	Chapter 20

Questions related to written presentation

73. Should I use a template for writing my report or thesis?	Chapter 12 (Use a Template), Chapter 22 (Create a Template File)
74. How do I type an equation?	Chapter 15 (Tips on Typing Equations)
75. How long should each section/chapter be?	Chapter 16 (Interesting Facts)
76. How do I format my report or thesis?	Chapter 22
77. What is the purpose of figures in a report or thesis?	Chapter 21 (Introduction paragraph, Purpose of Figures)
78. What are the requirements of photographs and images in a report or thesis?	Chapter 21 (Photographs and Images)
79. How do I make a schematic drawing?	Chapter 21 (Schematic Drawings)
80. How do I plot a graph?	Chapter 21 (Graphs and Charts)
81. When I insert a figure into my thesis, the figure does not stay in place. What can I do?	Chapter 21 (Tips for Positioning Figures, Images and Charts in Microsoft Office Word)
82. Should I use double-line spacing or single-line spacing in my report or thesis?	Chapter 22 (Follow the Guidelines)
83. How do I create a contents page?	Chapter 22 (Create a Template File)
84. What is a typical content structure for a report/thesis?	Chapter 22 (Layout of Report or Thesis)
85. How do I create a template file?	Chapter 22 (Create a Template File)

Questions related to oral presentation

86. What should I do to prepare for a project presentation?	Chapter 23 (Preparing for an Oral Presentation)
87. How long is a project presentation?	Chapter 23 (Presentation Time)
88. How many slides should I prepare for a presentation?	Chapter 23 (Presentation Time)

(Continued)

	Chapter (section) where questions are answered
89. What sections of my report or thesis should I include in a project presentation?	Chapter 23 (Presentation Format)
90. How do I structure a project presentation?	Chapter 23 (Presentation Format)
91. How should my presentation slides be prepared?	Chapter 23 (Presentation Slides)
92. How do I control my nervousness when presenting my report or thesis?	Chapter 24 (Managing Anxiety)
93. How can I speak clearly and confidently during my presentation?	Chapter 24 (Delivering Your Presentation)
94. How can I avoid making errors in pronunciation?	Chapter 24 (Delivering Your Presentation)
95. How do I use a laser pointer in a presentation?	Chapter 24 (Using Presentation Tools)
96. What should I do if I run out of presentation time?	Chapter 24 (Managing Your Time, Procedure for Finishing Quickly)
97. What should I do if I am asked to end my presentation in 2 minutes?	Chapter 24 (Tip: How to Find Your Conclusions Slide Instantly, Managing Your Time, Procedure for Finishing Quickly)
98. How to I prepare for questions at the end of my presentation?	Chapter 24 (Answering Questions)

Part I

Planning

Chapter 3

Choosing a research project

The sign of a good decision is the multiplicity of reasons for it.

Mary Doria Russell
Children of God, 1998

This chapter deals with a problem that students face when embarking on their first research project. How much importance should you place on the choice of a research project? What are the considerations in choosing a topic for your research project?

BEFORE YOU CHOOSE A RESEARCH PROJECT

Choosing a research project is an important decision and should be made carefully. A research project usually lasts an entire academic year for a final year undergraduate project (FYP), from 1 to 3 years for a master's project and from 3 to 6 years for a PhD project. Therefore, making a wrong choice means that you will have to live with the consequences of your choice for the rest of your candidature. Choosing the wrong project may have other serious consequences. For undergraduate students, it may affect the class of degree and their chances to progress to postgraduate studies. For postgraduate students (master's and PhD), it may be the difference between success and failure. Although postgraduate students have the option of changing the scope or even changing the topic later, this will mean that they have lost precious time during their candidature. It is, therefore, important to spend some time finding out as much as possible about a research topic before deciding on it.

Bear in mind too that universities have different practices regarding selection of research projects by their students. For undergraduate students and postgraduate students completing a degree by coursework, such practices can range from students choosing from a list of prescribed projects to students proposing a project of their own. Choosing a project from a pre-determined list has its advantages as well as disadvantages.

The advantages of choosing from a list of projects are that the objective and scope of the project have been generally spelt out and the resources needed for the project have been planned for. The disadvantages of choosing from a list of projects are that your choices are limited and you will likely have to compete with your peers for the projects you are interested in.

Postgraduate students doing a degree by research may have to propose a research project or work on a partially or wholly funded research project. Developing and submitting a project proposal involves more work (read *Chapter 5—Writing a research project proposal*). If the project is part of or the whole of a funded research project, the objective and scope are usually pre-defined. Such a project can be even more restrictive than an under-graduate project as the deliverables of the project have already been clearly identified. Your completed PhD thesis may also be different from the final report of the funded research project. Your thesis may cover either addi-tional scope not mentioned in the funded research project, or only a part of the scope of the funded research project.

BASIC CONSIDERATIONS FOR CHOOSING A RESEARCH PROJECT

It is advisable to think through the following questions before choosing your research project:

1. Which field in my discipline appeals to me? For example, a civil engi-neering student may consider fields such as structures, geotechnics, environment, hydraulics, hydrology, transportation, etc.; a mechani-cal engineering student may consider fields such as machines, manu-facturing, materials, robotics, mechanical and electrical services, fluids, thermodynamics, etc.; an electrical engineering student may consider fields such as electronics, power engineering, computer hard-ware, software engineering, mobile devices, integrated circuits, super-conductors, integrated circuit design, etc. It is also common these days to carry out an inter-disciplinary project, for example, in bioma-terials which is a combination of biology and mechanical engineering disciplines.
2. What type of research work would I like to conduct? Experimental, numerical, analytical, survey, software coding, design, etc.?
3. What are the resources available to me for carrying out the project? Resources include equipment, facilities, and funds. If equipment and facilities required are not available when you start the project, you may have to acquire or fabricate them, which yourself. This will likely affect the rate of progress of your research project.

4. How much do I already know about the research topic and what skills do I already have for researching it? You can answer these questions by examining the research topic carefully. Do you have some knowledge of the research topic? What is the extent of that knowledge? Do you have the skills required for the research method that will be used?

By answering the above questions, you will be able to arrive at a shortlist of projects for further consideration. For students who do not have a list of projects to choose from, you would have to first identify a field within your discipline, then narrow that down to the type of research work and, finally, the research method to arrive at a potential research topic.

Besides the preceding considerations, your final choice of a research project may involve non-technical considerations such as those indicated in advice for undergraduate and postgraduate students.

ADVICE FOR UNDERGRADUATE STUDENTS

After making a shortlist of the possible projects, ask yourself the following questions to shorten your list further:

1. What do I want to achieve from the project? Good grade? New skills? Potential topic for postgraduate studies? Improve prospects for employment after graduation? These goals may not be mutually exclusive. Sometimes it could be a combination of two or more of these goals. However, keep in mind that the purpose of an undergraduate project is for you to learn how to conduct research. Your performance in the undergraduate project may determine the class of degree you are awarded. Many universities require an honours undergraduate degree as one of the admission criteria for postgraduate studies.

2. What can I expect from my supervisor? Your project supervisor is very important to the success of your project. Besides being knowledgeable and experienced in the research area you are doing your project in, a supervisor needs to be interested in your research project and in your progress. The supervisor needs to be available and should set aside time for you when you ask to see him or her. The supervisor also needs to be firm when you are not making enough effort in your project and be willing to give you the 'push' you need. When you encounter problems in the project, a good supervisor will give you constructive feedback and guidance. However, a supervisor may intentionally leave you to work on a problem so that you can learn to develop independent problem-solving skills. When you lose focus,

the supervisor will provide direction. A supervisor will also be able to assess the contribution/value of your project.

Some students may expect their supervisors to solve all their problems but do remember that your supervisor is not the person doing your project, you are! The relationship between a supervisor and a student is a two-way street. You should update your supervisor regularly on your progress or problems that you encounter. When your supervisor gives you advice or guidance, you should act on it to ensure successful completion of the project. Therefore, working closely and maintaining regular contact with your supervisor is very important. In most cases, your supervisor will eventually become your mentor and, very likely, your referee in your job applications after graduation.

3. Will I have to work in a team for my research project? In some universities, undergraduate projects are carried out in teams of two and three students. You may or may not be able to choose your own team members. If you can choose your own team members, you will need to find those that share the same goals as you for the project. Preferably, team members should have similar academic abilities or complementary personalities to ensure the best outcome for the project. If you have to work with team members chosen for you, it is advisable for the team to meet at the start of the project and discuss the role and responsibilities of individual team members, lay the ground rules for collaborating with each other, and agree on the approach to be taken for completing the project.

By considering the preceding questions carefully, you will avoid having unrealistic expectations of the project. For example, if you want to work for an 'A' grade for the project, you should make this goal known to your project supervisor and ask about the amount of work and results expected of you to achieve this goal. Students who want to achieve a 'pass' grade for the project should likewise communicate this to the project supervisor so that the work scope and results are commensurate for a 'pass' grade. It is advisable to meet your prospective supervisor before choosing a project to reach a mutual understanding about what is expected.

ADVICE FOR POSTGRADUATE STUDENTS

Choosing a suitable research project is even more crucial for postgraduate students. It is likely that postgraduate students would have consciously or subconsciously weighed the considerations discussed in the previous section, as they would have had the experience of conducting an undergraduate research project before applying for a postgraduate studies programme. Relevant questions that postgraduate students should ask themselves are as follows:

1. What courses or actions do I have to take to gain the knowledge required for my research project?
2. Will the scope of the project be sufficient for the degree (master's or PhD)?
3. Will the research result in new knowledge or have impact on society? On the field?
4. What is the potential of my research field for my future career?

The preceding questions will help you to identify a clear personal goal for your candidature. When applying for postgraduate studies, it is important to be aware of the research strength of the university and that of your prospective supervisor. Contact your prospective supervisor as early as possible to discuss potential topics as well as support available for your studies. The support includes financial support, research facilities and, if relevant, access to data. You should also attempt to find out the supervisory style and history of your prospective supervisor by speaking to his past or current students. You can also look at the number of postgraduate students working under him, his track record in graduating students, the typical candidature period of the postgraduate students under him and the publications and theses of the prospective supervisor's past postgraduate students. You should also consider whether your prospective university has in place a system to handle cases where postgraduate students are not able to work with their supervisor. Such cases, though rare, are still a possibility. Finally, postgraduate students are expected to be more independent than undergraduate students and possess self-learning skills. Therefore, do not expect your supervisor to supervise you as closely as you were supervised for your undergraduate research project.

As in many things in life, you are unlikely to get everything you want with regard to choosing your research project, supervisor, or team members. If you cannot have what you want, you should try to like what you have. After all, an essential skill of an engineer is adaptability. Instead of wasting time feeling disappointed at not getting the project that you want, you will be better off reading up on the project and working out the project's requirements. Consulting your project supervisor about the project will very likely help you see the positive aspects of your project.

INTERESTING FACTS

A study by Harland et al. (2005) entitled 'Factors affecting student choice of the undergraduate research project: staff and student perceptions' showed very interesting results. The study was conducted through a survey questionnaire for students in Biomolecular Sciences at Liverpool John Moores University, UK. A total of 84 student responses and 20 staff responses were obtained. The survey revealed that students and staff broadly shared similar expectations for the project. Students rated interest in the subject as the most important factor in their choice of project. Students also rated 'the chance to use particular skills', 'the chance to extend knowledge in a familiar area', and 'the opportunity to learn more about an unfamiliar area or technique' highly, contrary to staff's expectation. A surprising outcome of the study was that about 60% of students were looking for a 'challenging project' while staff thought that only 16% of students would do such projects. The study also revealed that students see their project supervisor's role progressing during the project from one giving guidance on the project to a mentor providing academic advice and sometimes career guidance, and generally being a friend.

REFERENCE

Harland, J., Pitt, S., & Saunders, V. (2005). Factors affecting student choice of the undergraduate research project: Staff and student perceptions. *Bioscience Education 5*. doi:10.3108/beej.2005.05000004.

Chapter 4

Researching a topic

> There are some things which cannot be learned quickly, and time, which is all we have, must be paid for their acquiring.
>
> Ernest Hemingway
> *Death in the Afternoon, 1932*

The best way to learn about a topic is to read other researchers' publications. Reading other researchers' work is also an essential part of research. But before you can start reading other researchers' work, you will first need to find relevant publications. This process is called information searching and it is an essential research skill.

KEYWORDS

To search for relevant information in any sources, you will first need to identify the keywords that define your project. According to the Merriam-Webster dictionary, a keyword is 'a significant word from a title or document used especially as an index to content'. Keywords are used in all web searches and play an important role in the archiving of books and articles. Most journal articles list their keywords as these increase the chance of the articles being retrieved, read and cited by researchers. Three journal articles and their keywords are listed below:

Guo, H., Low, K. S., & Nguyen, H. A. (2011). Optimizing the localization of a wireless sensor network in real time based on a low-cost microcontroller. *IEEE Transactions on Industrial Electronics* 58(3): 741–749.
Keywords: Microcontrollers; particle swarm optimization; wireless sensor network.
Khong, P. W. (1999). Optimal design of laminates for maximum buckling resistance and minimum weight. *Journal of Composites Technology & Research* 21(1): 25–32.
Keywords: Laminates; fuzzy multi-optimization; finite strip analysis.

Leong, E. C., & Rahardjo, H. (2012). Two and three-dimensional slope stability reanalyses of Bukit Batok slope. *Computers and Geotechnics* 42: 81–88.

Keywords: Two-dimensional; three-dimensional; limit equilibrium; slope stability; residual soil; unsaturated; matric suction; groundwater table; rainfall; factor of safety.

If you read the title of an article and its list of keywords, you will see that some of the keywords appear in the article's title. Other keywords highlight the core content of the article.

Once you have a list of the keywords that define the research topic, you can use these keywords to search for relevant publications. Keywords and phrases can be combined using Boolean operators (AND, NOT, OR and NEAR) to limit, widen or define your search. Most online databases and Internet search engines support Boolean searches. Table 4.1 illustrates the use of Boolean operators to search for information.

SOURCES OF INFORMATION

The Internet is the most popular source of information. Be systematic when searching for information on the Internet and record your searches (see Chapter 8—Keeping Research Records). Start by writing down the

Table 4.1 Use of Boolean operators to search for information

Boolean operator	Example	Explanation
AND (alternate +)	Microcontrollers AND network, or Microcontrollers + network, or Microcontrollers network	AND combines two or more words in the search. By default, a search involving two words implies the AND Boolean operator is used. If a phrase rather than two separate words are to be searched, put the words in quotation marks, for example 'Microcontrollers network'. The more words or phrases join with AND, the fewer documents you will find.
NOT (alternate −)	Microcontrollers network—chips	Remove documents containing 'chips' from the list of documents containing 'Microcontrollers network'.
OR	Microcontrollers OR network	OR requires either of the words or phrases combined by OR to appear in the document. The more words or phrases you combined with OR, the more documents you will find.
NEAR	'Microcontrollers network' NEAR chips	Requires the word 'chips' to be found within the proximity of the phrase 'Microcontrollers network'. The number of words between the two search terms varies with the database and search engine.

keywords in the same order that they are typed including the Boolean operators to generate a list of search results. Next, write the URL of the web page
and the date that you accessed the web page as the information that you
captured may be changed at a later date. You can then either *copy and paste*
the information relevant to your topic onto a Word or similar type of word
processing document, or use software specially created for note-taking (e.g.
Diigo, Evernote, Microsoft Onenote, Springnote, Ubernotes, WebNotes).

Using the Internet to seek information, however, presents certain problems. First, you will have to sift through a vast amount of information to find
the information that is relevant to your research and, second, you will have
to determine the reliability of the information. The first problem may be
solved to a certain extent by using a list of more specific keywords or phrases
combined with the Boolean operator AND. But the second problem—determining the reliability of information—can only be resolved when you have
become more knowledgeable about the research topic. For example, your
search may produce a number of open access journal publications. Some of
these open access journal publications, however, may be of dubious quality. At the earlier stages of your project, you may find it difficult to judge
the quality of a journal article. Examples of dubious journals to avoid can
be found at a website, http://scholarlyoa.com/, created by Jeffery Beal, an
academic librarian who maintains a database of such journals.

A more reliable way of seeking information on your topic is to use your
institution's library resources. Books and published theses are excellent
publications to start your research on a topic. Besides hardcopy publications, check whether your institution's library subscribes to electronic
databases. You can search electronic databases for articles published in
conferences and journals, as well as for patents. Engineering databases such
as *Engineering Village* (Engineering Index) by Elsevier and *Web of Science*
by Thomson Reuters provide a consolidated electronic resource for you to
search for published information. There are also specialised databases for
specific fields though *Engineering Village* and *Web of Science* should be
your primary databases for your search. You can search the databases by
using the same keywords you used for Internet searches. The information
provided by the databases includes the title, abstract, source of the article,
and citation data.

TYPES OF INFORMATION

Once you have obtained potentially relevant information on your research
topic, you will need to evaluate it critically. The following questions will
guide you in your evaluation:

1. How relevant is the information to your research topic? You may find
   that only part of the information found is relevant.

2. If the information is relevant, what is its significance? You may want to categorise this information under research method, materials and methods or results and discussion.
3. If the information is related to research method, is the research method clearly explained? Is it the only research method? Why did the author choose this research method? Is the research method grounded on theory?
4. If the information is related to materials and methods, what were the materials and methods used? How were the experiments designed? How were the measurements made? How reliable and accurate are the measurements?
5. If the information is related to results and discussions, how did the author analyse and interpret the data? Is there an alternative method of analysis? Is the interpretation sound? Does the conclusions reflect the data and analysis?
6. How does the information contribute to your understanding of the research topic? What are the strengths and limitations of the information?
7. Is there any information gap in the research that you can fill?

As you research a topic, you will come across some papers that have high citations, that is, the papers have been referenced (cited) by many researchers. Such papers are key papers that you should make a note of, particularly if they are directly related to your research topic.

More information on researching a topic can be found in *Chapter 13—Writing introduction and literature review*, and *Chapter 19—Using sources and avoiding plagiarism*.

INTERESTING FACTS

Engineering Village originated from *Engineering Index* (Ei). Ei was created by Dr. John Butler Johnson, a professor of civil engineering at Washington University in St. Louis, Missouri in 1884 out of a need to gain knowledge of engineering literature. Ei consists of an index and the abstracts. In 1918, Ei was acquired by the American Society of Mechanical Engineers. In 1967, the first Ei electronic bulletin, 'Current Information Tape for Engineers' (CITE), was produced. In 1969, CITE evolved into COMPENDEX (COMPuterized ENgineering inDEX). In 1995, EngineeringVillage.com was launched. In 1998, Elsevier purchased Ei. Currently *Engineering Village* covers 12 databases containing patents, monographs, books and dissertations.

Web of Science by Thomson Reuters evolved from the Science Citation Index (SCI) created by Dr. Eugene Garfield. Dr. Garfield founded the Institute for Scientific Information (ISI) located in Philadelphia, Pennsylvania, in 1955. He is responsible for creating Current Contents, SCI, the Journal Citation Reports, Index Chemicus and other citation databases. Currently, *Web of Science* consists of seven databases and boasts extensive coverage of more than 250 disciplines in the sciences, social sciences, arts and humanities.

Chapter 5

Writing a research project proposal

> Imagination is more important than knowledge. For knowledge is limited to all that we now know and understand, while imagination embraces the entire world, and all there ever will be to know and understand.
>
> Albert Einstein (1879–1955)

Proposals in engineering and the technical sciences require not just knowledge of the field but also imagination to apply that knowledge to solve problems. This chapter aims to help you write a proposal for a research project. A research proposal is a standard requirement for acceptance into most postgraduate programmes. (However, if you are applying for a specific pre-defined research project in an engineering undergraduate programme, you may not need to submit a project proposal.)

In a good research proposal, you will need to demonstrate that you are

1. Capable of independent and critical thinking and analysis
2. Capable of communicating your ideas clearly

PURPOSE OF A RESEARCH PROPOSAL

The proposal is intended to show your supervisor that you have a clear idea of previous work in your chosen area of research, the research problem and the methods you plan to use to solve the problem.

STRUCTURE OF A RESEARCH PROPOSAL

The structure and length of your research proposal will vary depending on the requirements of your institution or department so the first step is to find out departmental guidelines and requirements.

Examples of requirements for a research proposal are the guides provided by the University of Edinburgh (*How to write a good postgraduate research proposal*, 2010), and the University of Melbourne (*How to write a research proposal: A guide for science and engineering students, n.d.*). Typically, a research proposal comprises the following elements:

Title: A well-crafted title is essential for two reasons. First, it acts as the gateway to the content of your research proposal and, therefore, should tell the reader immediately what the proposal is about. Second, it is the first part of your research proposal that is read by the reviewer or research committee, and it is essential that you make a positive first impression.

Obviously, the title you come up with at this stage is only a working title. It is unlikely to be the final title of your completed research project as this will often depend on your results. Here are three guidelines for writing an effective title:

1. Indicate clearly the content and focus of the research project.
2. Make it clear and concise. The primary function of a title is to provide a precise summary of the paper's content so avoid unnecessary details. A good title should be no more than 15 to 20 words.
3. Make it descriptive; include keywords that describe the proposal.

A common practice is to start with the main overall topic and use a colon to separate the topic from the focus as in the following example:

Storm water harvesting: managing the hazards of surface water pollution by run-off.

Abstract (for longer proposals): In the abstract, summarise the gist of your proposed research project in a paragraph of about 150–200 words. State the purpose and motivation for the study, a statement of the problem, the data collection methodology and analysis, and the significant results and implications of the research proposed.

Introduction: This section provides background information for the research (i.e. the problem being addressed). Information relating to your research question(s) or hypothesis/hypotheses is typically organised from general to specific.

The background information is provided in the form of a literature review that helps you set the context for your research to help the reader understand the research questions and objectives. Choose key research papers and explain clearly how your research will either fill a knowledge gap or follow from previous research (refer to *Chapter 13—Writing the Introduction and Literature Review* for practical information on how to conduct and write a literature review).

Possible format for an Introduction:

- Introduce the area of research
- Review key research papers
- Identify any gap in knowledge or questions that needs to be answered
- Your hypotheses or research objectives
- Scope of your research project

Research Problem: State the primary problem you are trying to solve. It may be a research question(s) or hypothesis/hypotheses stated in a few concise sentences that express clearly the focus of your project and its scope.

Methodology: Include a description and rationale for the methods of data collection and analysis, and the materials used. Provide just enough information for the experiments and data collection to be replicated by someone else, and nothing more.

Typically, this section uses subheadings (i.e. *Subjects, Instrumentation, Data Collection, Methods of Analysis etc.*) and is written using the future tense, for example, 'The research will initially examine water treatment processes in ...'

In deciding on the subheadings to use, think of the kind of research you are proposing:

- *Experimental*—equipment, materials, method
- *Modelling*—assumptions, mathematical tools, method
- *Computational*—inputs, computational tools, method

Timetable/Schedule: List the stages of the research project in timeline or tabular format and the deadlines for completion of these stages or tasks.

Report/Thesis Organisation (if applicable): Outline the proposed chapters of the report or thesis and describe the content of each chapter in a few sentences.

Significance and Contribution of the Study: Conclude the proposal with a paragraph on the importance of your research and state how it will contribute to knowledge and understanding of certain issues. Relate the expected outcomes of your research to the objectives stated in the Introduction, so that the significance of your study and the contribution to knowledge is apparent. However, do not exaggerate the importance of the contribution your study will make.

List of References: List all the sources cited in your research proposal by using a referencing format appropriate to your institution or department. Do not list references that are not referred to in your proposal.

An example of a research project proposal is shown in Writing Guidelines 5.1.

Writing Guidelines 5.1 A sample research project proposal

Abstract/Summary (optional)

See Chapter 17:
Abstract

Background/Literature Review

Various aspects of civil engineering require information that is
not only accurate, but also up-to-date. In geotechnical
engineering, for example, it is important to have detailed
knowledge of the soil conditions and terrain. While the soils
can be investigated through point sampling, studying the macro
features of a terrain would be more challenging as the area of
study would be much larger. Conventional methods of point
sampling and mapping to a large area would be very
expensive and time consuming. The recent advances and
availability of satellite platforms such as World-View,
QuickBird, IKONOS, LANDSAT, AQUA, TERRA and locally
owned XSAT for civilian usage have now allowed quantitative
and qualitative evaluations of any large area via satellite
imagery.

See Chapter 13:
*Writing the
Introduction and
Literature Review*

Satellite imagery has been utilised in various disciplines,
including the estimation of evapotranspiration (El Tahir et al.,
2012) as well as the mapping and quantification of land area
and cover types (Kamaruzaman and Hasmadi, 2009). These
applications use satellite images that were captured over
longer time intervals. Applications in geotechnical engineering
require satellite images to be acquired over shorter time
intervals. This is currently being realised with the launching of
micro- and nano-satellites where images can be acquired at
time intervals of days rather than weeks or months
(Nakasuka, 2013). This provides the opportunity to apply
satellite imagery to more rapid change events like flooding,
landslides, earthquakes and so on.

Research Objective

The objective of this research is to use satellite imagery to show
correlation among different variables for landslides. These
variables include soil water content, temperature distribution,
precipitation, land cover and topography.

Methodology

Satellite images containing landslide sites with ground truth
data will be collated. Information on the different variables
will also be obtained from satellite data. It is envisaged
that the data will be obtained from several satellites. The
information obtained will be put into a database for
analysis and establishment of correlations. In addition,
deterministic analysis for individual landslides will be
conducted to understand the underlying mechanism of the
landslide.

See Chapter 7:
*Research
Methodology and
Research Methods*;
Chapter 14:
*Writing the
Materials and
Methods Section*

(Continued)

Writing Guidelines 5.1 (Continued) A sample research project proposal

Timetable/Schedule

The collation of information will take about 2–3 years. However, the information collated by the end of the first year will suffice for the start of data analysis. It is expected that the development of correlations will take a further 2 years. Overall, the project is expected to be completed within 4 years.

See Chapter 6:
Planning a Research Project

Conclusion

With a better understanding of the variables contributing to landslides, it is possible to develop an early warning system for landslide prone areas, more *reliable* landslide risk maps as well as *more effective* policies for sustainable development of hilly terrain.

Resources

The university has already the software and hardware for the processing of satellite images. As far as possible, the project will endeavour to use satellite data that are publicly available. However, the project may require the purchase of additional high-resolution satellite images and satellite data for in-depth study of some of the landslides.

References

El Haj El Tahir, M., Wenzhong W., Xu C.Y., Youjing, Z., & Singh V. P. (2012). Comparison of methods for estimation of regional actual evapotranspiration in data scarce regions: The Blue Nile, eastern Sudan. *Journal of Hydrologic Engineering* 17: 578–589.
Kamaruzaman Hj, J., & Mohd Hasmadi, I. (2009). Mapping and quantification of land area and cover types with Landsat™ in Carey Island, Selangor, Malaysia. *Modern Applied Science* 3(1): 42–50.
Nakasuka, S. (2013). Current Status and Future Vision of Hodoyoshi Microsatellites—Systems for Quick and Affordable Space Utilizations. Proceedings of the 5th Nano-Satellite Symposium, November 20–22, 2013, Tokyo, Japan.

Source: Lim, B. J. M., *PhD project proposal* (unpublished), Nanyang Technological University, Singapore, 2013.

OTHER CONSIDERATIONS

1. *Resources:* When writing a research project proposal, you will need to consider the resources (equipment, test materials etc.) available for your project. If not, you will need to consider the possibilities of acquiring such resources during your project.
2. *Ethical Considerations:* Research done in a university setting usually requires Institutional Research Board (IRB) approval. This means that your research has to be approved by an ethics committee to make sure

you comply with the rules and expectations with which the research should be conducted. Project proposals must therefore include potential issues raised by the conduct of the research and how these will be addressed should they occur. This is particularly important if your research is deemed 'high-risk', that is if it involves people, animals, or sensitive materials. Find out what the ethical approval system is in your prospective institution. Your supervisor will be able to provide you with this information.

TIPS ON WRITING STYLE AND LANGUAGE IN A RESEARCH PROPOSAL

1. Follow the three C's rule:
   Clear: Is what you have written intelligible and are your ideas clearly articulated?
   Concise: Have you written your proposal in a succinct and focused way?
   Coherent: Are the sections of your proposal clearly linked so that it is clear to the reader what you want to do, why you want to do it and how you will do it?
2. Revise and edit your writing thoroughly:
   Poor grammar and inappropriate style distract your reader and compromise your credibility as a researcher. Use spell check and grammar check applications. (Refer to *Chapter 10—Grammar, Punctuation and Word Usage Guide* and *Chapter 20—Revising and Editing* for help with grammar and revising.)
3. Use transitions:
   Signal to the reader as you move through your text by using transition words and expressions such as *however, following this, in contrast, consequently* and so on.
4. Avoid overly hesitant or tentative language:
   Sound confident and sure about the work that you are proposing to do. So avoid excessive use of expressions such as *it seems that…, it is hoped that…, it might be possible…, perhaps* and so on.

CHECKLIST FOR WRITING A RESEARCH PROPOSAL

Does your proposal include

☐ A critical discussion of previous research in your area?
☐ A clear statement of your hypothesis/hypotheses or objective(s)?
☐ The methodology for your research and the expected stages?
☐ Facilities, resources, laboratory equipment needed?

INTERESTING FACTS

Writing research project proposals is a normal activity in academia. Such research project proposals are often for research grant applications. However, the success rate of obtaining a grant is about 20% (American Society for Engineering Education, n.d.), that is one successful application in five applications. The success rate for the individual may be as low as one successful research project proposal for every 10 or 20 written research project proposals.

REFERENCES

American Society for Engineering Education (n.d.). *Tips for a winning research proposal*. Retrieved from: http://www.asee-prism.org/tips-for-a-winning-research-proposal/.

Lim, B. J. M. (2013). *PhD project proposal* (unpublished), Nanyang Technological University, Singapore.

The University of Edinburgh. (2010). *How to write a good postgraduate research proposal*. Retrieved from: http://www.ed.ac.uk/polopoly_fs/1.58205!/fileManager/HowToWriteProposal.pdf.

The University of Melbourne. (n.d). *How to write a research proposal: A guide for science and engineering students*. Retrieved from: http://services.unimelb.edu.au/__data/assets/pdf_file/0006/471273/Writing_a_research_proposal_Science_Engineering_Update_051112.pdf.

Chapter 6

Planning a research project

He who fails to plan is planning to fail.

Benjamin Franklin (1706–1790)

A research project, unlike a course, has no class schedule. In other words, a research project is an unstructured course that requires the student to organise and manage his time for conducting the research.

PHASES OF A RESEARCH PROJECT

The way most students approach their research project can be described using the six phases identified by Holland (2001) for project management: (1) enthusiasm, (2) disillusionment, (3) panic and hysteria, (4) search for the guilty, (5) punishment of the innocent and (6) praise and honour for the non-participants.

The *enthusiasm phase* is experienced at the start of the research project. Students get excited over the prospect of conducting some 'groundbreaking' research. Driven by euphoria, they devote all their waking hours including cutting classes to conduct research. However, the enthusiasm dies down once they discover that research work is tedious and the progress is slow. They start to realise that research work is fraught with problems that they cannot solve easily. This leads to the *disillusionment phase* where they start to lose interest and plod along indifferent to the quality of the research, hoping that the research project will soon end. As the end of the candidature approaches, they suddenly enter the *panic and hysteria phase* as they realise that they have not done enough work. They start spending many sleepless nights to complete the research work but discover that they are unable to meet the deadline. Soon, they look for excuses for not being able to complete the research work (*search for the guilty phase*) and start to blame others for their failures (*punishment of the innocent phase*). Finally, they manage to complete the thesis for submission. On their acknowledgements page, they thank their friends and family members, who have no idea about the research project, for 'contributing' to the successful completion of their theses (*praise and honour for the non-participants phase*).

The preceding scenario need not happen to you with proper time management, planning and realistic expectations. A research project usually stretches over one academic year for an undergraduate project and two or more years for a postgraduate project. It is more akin to a marathon race than a short sprint. Therefore, you will want to plan your time for your research project carefully to ensure that you sustain interest and achieve the desired outcomes. Your planning should cover the entire duration of the project. Only after you have drawn up a project schedule will you have more realistic expectations of yourself and of your research project.

PROJECT MANAGEMENT

When you are planning your research project, you can apply project management principles. First, from the objectives of the project, work out the scope that will enable you to fulfil these objectives. From the scope, identify the specific tasks or activities that need to be carried out and define the work within each activity. The activities identified, in the first instance, can be very broad such as conducting literature search, setting up experiments, conducting experiments, analyzing test results and report writing. When you have done more work, you can make the activities more specific. For example, you can then identify the topics in your literature search and break down your literature search into several sub-activities according to topics instead of as a single activity.

Once the activities are identified, you can start to estimate the amount of time that each activity will take. It may not be easy to estimate the time needed for an activity as an activity is subject to the triple constraints of time, resources and scope. Resources can be funding or equipment. Scope refers to the amount of work involved. If resources are fixed, the time (duration) for the scope of work is fixed. However, time for an activity can be reduced if more resources are injected into the activity. For the duration of an activity that deals with a research method that you have no prior experience of, you can consult past students or your supervisor for help. Once you have worked out the activities and the time involved, you can identify which activity depends on the start or completion of another activity before it can begin.

After the concerns above are addressed, you can then work out the critical path of your research schedule, which is the string of research activities that takes the longest time. The activities can now be put together into a schedule that is commonly known as a Gantt chart. For research work, you can plan your activities in terms of a week, fortnight, month or quarter depending on the duration of the research project and the frequency with which you meet your supervisor. There is dedicated software for creating a

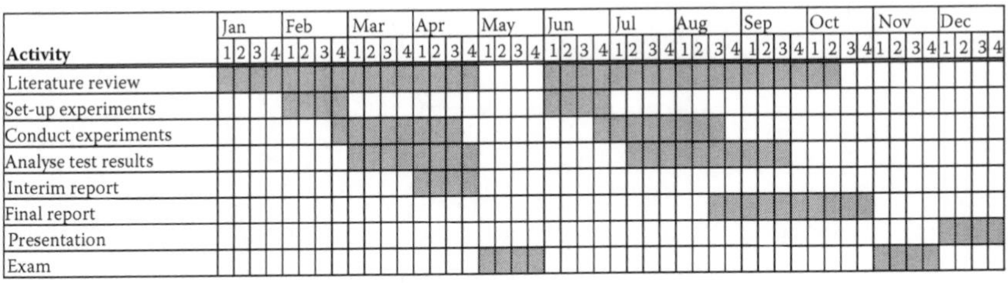

Activity	Jan				Feb				Mar				Apr				May				Jun				Jul				Aug				Sep				Oct				Nov				Dec			
	1	2	3	4	1	2	3	4	1	2	3	4	1	2	3	4	1	2	3	4	1	2	3	4	1	2	3	4	1	2	3	4	1	2	3	4	1	2	3	4	1	2	3	4	1	2	3	4
Literature review	▓	▓	▓	▓	▓	▓	▓	▓	▓																				▓	▓	▓	▓	▓	▓	▓	▓												
Set-up experiments					▓	▓	▓	▓	▓																																							
Conduct experiments									▓	▓	▓	▓	▓								▓	▓	▓	▓	▓	▓							▓	▓	▓	▓	▓									
Analyse test results									▓	▓	▓	▓	▓														▓	▓	▓	▓	▓	▓	▓	▓	▓	▓											
Interim report													▓	▓	▓	▓	▓																															
Final report																																					▓	▓	▓	▓	▓	▓	▓					
Presentation																																															▓	▓
Exam																	▓	▓																										▓	▓			

Figure 6.1 Example of a Gantt chart.

Gantt chart such as GanttProject, SmartSheet, SmartDraw and Microsoft Project. However, a simple Gantt chart can be created using Microsoft Office Excel or using a table in Microsoft Office Word. Figure 6.1 shows an example of a simple Gantt chart for a 1-year research project created using Microsoft Office Excel (see the section *Tips on Making a Gantt Chart* in this chapter). Note that in Figure 6.1, each month is divided into 4 weeks, some of the research activities occur concurrently and examination, a non-research activity, is included to explain breaks in the research activities. In planning, you should also cater for contingencies and allocate more time to activities that have a higher level of uncertainty or a greater certainty of failure. Many students also do not allocate enough time for report writing at the end of the project. The report or thesis is an important document that informs others of your work and not allocating enough time may jeopardise the outcome of your work. The Gantt chart is flexible and can be updated as you progress in your research project. It is a tool to help you manage your time so that you can complete the research project on time and on target.

BE ACCOUNTABLE

Once you have worked out the schedule of the research project, you will need to stick to it just as you did with the curriculum timetable for the courses that you take. The difference between the curriculum timetable and the research schedule is that the curriculum timetable is in term of hours whereas the research schedule is in term of weeks or of longer duration. Therefore, you need to allocate your time to the research activities on a daily basis to match the research schedule. If you are an undergraduate student, a good rule of thumb to use when allocating time for research activities is to set aside the same number of hours per week for the research as you would spend on an equally weighted course. For example, if the academic weightage of your final year project is equivalent to three courses and if each course requires you to attend 3 hours of classes, the *minimum* number of hours that you should put aside each week for research is 9 hours.

Meeting your supervisor regularly is a good motivation for you to keep to your research schedule. However, your supervisor may not be able to meet you as regularly as you would like. If that is the case, you should summarise your progress (weekly, fortnightly or monthly) into a short report (half page to one page) to update your supervisor. A sure sign that you are falling behind the research schedule is when you find that you have nothing to update your supervisor on and then make excuses to avoid meeting your supervisor.

BE FOCUSED ON YOUR OBJECTIVES

It is important to always keep the objectives of the research project in focus. It is quite common to 'stray' away from these objectives while trying to solve a problem. To find the solution, you may wander on to another topic and before long, you find yourself working on an entirely different topic. You will know if you are 'straying' away from the project by asking yourself the following questions:

1. Is there a connection between what I am doing now and my project's objectives?
2. Is what I am doing now important to my research project?
3. How will what I am doing now contribute to my research?

If your answers to the preceding three questions are 'No', then you are very likely straying from your project's objectives. You should stop what you are doing, and re-focus on your project's objectives.

BE REALISTIC

Thomas Edison, best known for his invention of the electric light bulb, made the following observation: 'Through all the years of experimenting and research, I never once made a discovery. I start where the last man left off ... All my work was deductive, and the results I achieved were those of invention pure and simple.' This quotation points out that most great discoveries were not made by chance, but through many hours spent on research and often based on the pioneering work of previous researchers.

Remember that most research projects only lead to incremental progress in knowledge and that only a few will lead to a major discovery. However, you should not despair as your work may still contribute to the next big discovery. The degree of new and innovative knowledge expected in your research depends on the type of degree you are pursuing. If you are pursuing a PhD degree, demonstration of new and innovative knowledge in your thesis is a pre-requisite for the award of the degree.

KEEP INTERESTED

Maintaining interest in a research topic requires discipline and self-motivation. Many researchers work on just one area over many years to become an expert in that area. The following are suggestions on how to maintain interest during your research:

1. *Read in-depth on your topic.* As you read in-depth, you will discover new things about your topic that will keep you challenged and make your project interesting.
2. *Meet your supervisor regularly.* Discussing your research with your supervisor on a regular basis will deepen your understanding of the research topic. Sometimes, by talking through the problems that you have encountered, you discover solutions that you would not have discovered if you had kept the problems to yourself.
3. *Switch among your research activities.* Sometimes, research activities are scheduled concurrently. When you find yourself unable to make progress in one research activity, switching to another research activity can make you productive again.
4. *Work regular hours.* Working unreasonably long and irregular hours can be detrimental to your physical as well as psychological health. Sleep deprivation will make you more prone to accidents and mishaps. Be mindful of this if your research requires you to work in the laboratory or to operate machines. With insufficient rest, you will also not be able to maintain interest in your research. Make sure you get the required number of hours of sleep for proper functioning—7 to 9 hours is recommended by the National Sleep Foundation (2015).
5. *Have a support group.* The support group can include fellow students and friends. You can check your progress with fellow students to see if you are progressing satisfactorily. This sometimes provides you with the motivation to persevere through difficult research activities. Fellow students and friends can support you by sharing your frustrations and problems. When you share your problems, you will find that you are better able to cope emotionally and psychologically with the demands of your research.
6. *Schedule breaks.* It is a good idea to plan breaks in your research schedule. The breaks allow you to evaluate your research progress and schedule. If necessary, realign some of your research objectives and then update the research schedule.

TIPS ON MAKING A GANTT CHART

A simple Gantt chart can be made using Microsoft Office Excel or using a table in Microsoft Office Word. Use the following steps in creating a Gantt chart:

1. Identify and list the main research activities in your project.
2. Break down the main research activities into smaller sub-research activities, if possible.
3. Estimate the time required for each main research or sub-research activity.
4. Identify the order of your research activities and the research activities that are to be performed concurrently.
5. If you are using Microsoft Office Excel, open a new spreadsheet. In the second row, enter 'Activity' in the first column and in the subsequent columns label the week numbers (1–4) for each month for the duration of your project as shown in Figure 6.1. Re-size the columns for the week number such that they are just wide enough to contain the week number as shown in Figure 6.1. You can do this by first selecting all the columns containing the week numbers and then select *Cells, Format* and then *Column width*. Enter the column width by entering the width in the pop-up dialog box (see Figure 6.2). Merge the first and second row in the first column (see Figure 6.3) so that 'Activity' appears as shown in Figure 6.1. Next, merge the first row of the next four columns for each of the months and enter the month's name as

(a)   (b)

Figure 6.2 Adjusting column width in Microsoft Office Excel. (a) Select *Format* in *Cell Size* group and select *Column Width*. (b) In the *Column Width* pop-up dialog box, enter the column width to adjust.

shown in Figure 6.1. Repeat for the subsequent months until the end of the project duration.

6. If you are using Microsoft Office Word, open a new Word document and select *Page Layout* and then *Orientation* to set *Landscape* mode. Select *Insert*, *Tables* and create a table with the required number of columns (total number of weeks + 1) and the required number of rows (total research activities, main and sub, + 1). In the second row, enter 'Activity' in the first column and in the subsequent columns label the week number (1–4) for each month for the duration of your project as shown in Figure 6.1. Re-size the columns for the week number so that they are just wide enough to contain the week number as shown in Figure 6.1. You can do this by first selecting all the columns containing the week numbers. Select *Table Tools*, *Layout* and then adjusting the column width in the *Cell Size* group (see Figure 6.4). Merge the first and second rows in the first column by using *Merge Cells* under *Merge* group (see Figure 6.5) so that 'Activity' appears as shown in Figure 6.1. Next, merge the first row of the next four columns for each of the

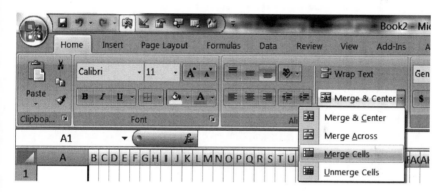

Figure 6.3 Merging cells in Microsoft Office Excel.

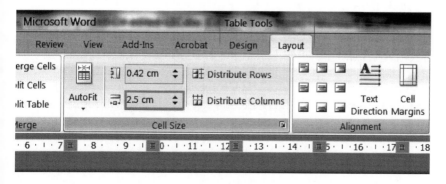

Figure 6.4 Adjusting width of a column in a table in Microsoft Office Word.

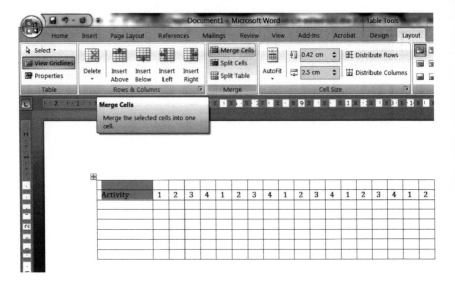

Figure 6.5 Merging cells in a table in Microsoft Office Word.

months and enter the month's name as shown in Figure 6.1. Repeat for the subsequent months until the end of the project's duration.

7. Enter the main and sub-research activities in the Activity column in the order that they should be performed.

8. For each main and sub-research activity, identify the start week number of that activity. Next, shade or mark with the letter 'x' the columns corresponding to the duration for that activity. To update your progress on the Gantt chart, use a different shade or letter for the appropriate week number columns.

INTERESTING FACTS

The Gantt chart was first developed by Karol Adamiecki, a Polish engineer, in 1896. However, it is widely known as Gantt chart after Henry Gantt, an American engineer, who independently developed the chart in 1903 as a management tool which became widely popular in Western countries.

REFERENCES

Holland, W. E. (2001). *Red Zone Management: Changing the Rules for Pivotal Times*. Chicago: Dearborn Trade, Kaplan.

National Sleep Foundation (2015). *How Much Sleep Do We Really Need*. Retrieved from http://sleepfoundation.org/how-sleep-works/how-much-sleep-do-we-really-need

Chapter 7

Research methodology and research methods

> There are two ways of acquiring knowledge, one through reason, the other by experiment.
>
> Roger Bacon
> *On Experimental Science, 1268*

Research methodology and research methods are often confused and are perceived as meaning the same thing even by researchers. In fact, *research method* refers to a specific process of conducting the research work, for example, experimental research, quasi-experimental research and survey research. *Research methodology*, on the other hand, explains how research is carried out in order to achieve the objectives of the research.

It is not possible to explain research methodology and research methods in detail in a single chapter or even a book. However, the brief account of research methods given should be sufficient to start you on your research journey and to guide you in looking for more in-depth information on a specific research method.

RESEARCH METHODOLOGY

Research methodologies can be classified differently. Most people will think of research methodology in relation to the type of research, which is either qualitative or quantitative. Qualitative research is associated with the use of words, sounds, images and objects as the data for research and the interpretation of the data can be subjective. In contrast, quantitative research is associated with numerical data that are obtained from experiments, surveys or past records and the interpretation of data is regarded as more objective. According to Borrego et al. (2009), theory is applied early in quantitative research to identify a hypothesis or hypotheses and to select suitable methods of data collection whereas theory is applied much later in qualitative research to interpret the findings. The differences between quantitative and qualitative research are illustrated in Table 7.1.

Most research in engineering is quantitative. Therefore, engineering students are more familiar with quantitative research than qualitative research. In quantitative research, engineering students usually design experiments to collect data with procedures that can be standardised and repeated. Unlike quantitative research, qualitative research requires a more careful design of the research method so that meaningful data are collected to present a consistent and coherent picture to generate the hypothesis. Only in research involving case studies with social contexts or surveys to collect information is there a need to apply some qualitative methods of research. Most engineering students do not pay much attention to research methodology or research methods as they tend to use the same research methodology and research methods gleaned from the literature or used by past students. The weakness with this practice is that they may be propagating the deficiencies of earlier research and limit their research achievements.

Table 7.1 Differences between quantitative and qualitative research

Quantitative	Qualitative
Objective is to test hypotheses that the researcher generates.	Objective is to discover and encapsulate meanings once the researcher understands the data.
Concepts are in the form of distinct variables.	Concepts tend to be in the form of themes, motifs, generalisations and taxonomies. However, the objective is still to generate concepts.
Measures are systematically created before data collection and are standardised as far as possible; for example measures of job satisfaction.	Measures are more specific and may be specific to the individual setting or researcher; for example a specific scheme of values.
Data are in the form of numbers from precise measurement.	Data are in the form of words from documents, observations and transcripts. However, quantification is still used in qualitative research.
Theory is largely causal and is deductive.	Theory can be causal or non-causal and is often inductive.
Procedures are standard, and replication is assumed.	Research procedures are particular, and replication is difficult.
Analysis proceeds by using statistics, tables or charts and discussing how they relate to hypotheses.	Analysis proceeds by extracting themes or generalisations from evidence and organising data to present a coherent, consistent picture. These generalisations can then be used to generate hypotheses.

Source: Neuman, W. L., *Social Research Methods* (2nd ed.), Allyn & Bacon, Boston, MA, 1994.

RESEARCH METHODS

Research should be based on the scientific method. The scientific method is generally described as consisting of six to eight stages. Hitchcock and Hughes (1995) described the scientific method as consisting of eight stages: (1) state the problem; (2) design the experiment, collect samples and identify the variables; (3) observe correlations and identify patterns; (4) form hypotheses to explain regularities; (5) test explanations and predictions (falsifiability); (6) develop laws or reject the hypothesis; (7) make generalisations; and (8) form new theories.

In this chapter, research methods are discussed in terms of experimental, quasi-experimental and survey research from the engineering perspective.

Experimental research

Experimental research involves research to determine causal relationships or testing cause–effect hypothesis (Dane, 1990; Edmonds and Kennedy, 2013). A cause–effect relationship is one where the occurrence of an event (the cause) causes the occurrence of another event (the effect). To establish a cause–effect relationship, three conditions must be present (Cook and Campbell, 1979):

1. When the cause occurs, the effect must follow. This is known as *temporal precedence*.
2. Whenever the cause occurs, the effect must also occur and the stronger the cause, the stronger will be the effect. This is known as *covariation*.
3. The effect must be due to the cause and no other causes can explain the effect.

Experiments can be conducted using standard or non-standard procedures. Standard procedures refer to procedures that have been standardised in national or international standards such as ISO, American Standards of Testing Materials or British Standards for the measurement of a test variable. Non-standard procedures refer to an experiment that is specially designed to measure a test variable. When using non-standard procedures, development of the non-standard procedures is an important part of the research.

The following are important considerations in experimental research:

1. *Dependent and independent variables* are present in all experiments. These variables need to be first identified so that the experiments can be designed to show the cause–effect relationships between the independent and the dependent variables. By systematically varying one dependent variable while keeping the other dependent variables

constant, the cause–effect of the dependent variable can be determined in a parametric study.

2. *Sampling method and sampling bias* need to be established when conducting tests on selected samples from a population. There is a need to show that the sampling method provides a sufficient number of unbiased samples, representative of the population, for testing. There are many different sampling methods and these can be grouped into probability and non-probability sampling methods. Probability sampling methods are more appropriate for quantitative research.

   Probability sampling methods include simple random sampling, cluster sampling, stratified sampling, systematic sampling and multistage sampling. The minimum sample size in probability sampling can be computed for based on whether the population size is known or unknown (Cochran, 1963; Krejcie and Morgan, 1970). When it is not possible to follow the recommended minimum sample size, 30 samples should be used. Jed Campbell explains this recommendation in 'Why is 30 the "Magic Number" for Sample Size?' (http://www.jedcampbell.com/). Non-probability sampling methods are more suitable for qualitative research and the sample size is usually small.

   Non-probability sampling methods include convenience sampling, maximum variation sampling, snowball sampling, typical case sampling and theoretical sampling.

   For more details on sampling methods, refer to Levy and Lemeshow (2009) and VanderStroep and Johnson (2010).

3. A *control experiment* is needed to establish that the effect is due to the cause and no other causes can explain the effect. By comparing the measurements from the control experiment with the other experiments, the reliability of the results can be increased.

4. *Repeatability and reproducibility* must be demonstrated in experimental research. Repeatability refers to obtaining the same experimental results under the same test conditions. Reproducibility refers to obtaining the same experimental results when the test is conducted by another person or at another time.

5. *Accuracy and precision* refers to the reliability of the measurement. Accuracy refers to the measurement error between the measured value and the actual value. Precision refers to the closeness of two measurements. Good experimental data should be both accurate and precise.

6. *Estimation of uncertainty* is required as all measurements involve some error, which is technically known as the uncertainty. An experiment should provide some measurement of the uncertainty.

7. *Ethical issues* are involved when dealing with living subjects and sensitive data or when the outcome of the research leads to harm to humans, animals or the environment.

Quasi-experimental research

Quasi-experimental research refers to non-randomised intervention studies. Another term used for quasi-experimental research is non-experimental research (Edmonds and Kennedy, 2013). Quasi-experimental research in engineering includes development or design of a product or system where the product can be software, material, equipment or a process. Examples of quasi-experimental research include research in software design, transportation, construction management, manufacturing engineering, industrial engineering and system engineering. In quasi-experimental research, there may be some cause and effect experiments but the emphasis is on changes to the developmental or design process. Quasi-experiment designs include pretest–posttest, time series and case studies. Quasi-experimental research tends to suffer from two types of errors: (1) specification errors that refer to cases when relevant dependent variables are not identified nor included in the analysis; and (2) self-selection errors that arise because of the non-random assignment. In other words, the assignment of the independent variable to specific categories is related to the status of the dependent variable.

Survey research

Survey research involves analysing information obtained from questionnaires or interviews. The survey is usually conducted on a small sample of the population. The information obtained from survey research comprises facts, opinions and behaviours (Dane, 1990). A fact is an observable phenomenon or characteristic that can be verified independently. An opinion refers to the belief of the survey's subject. The belief can be a preference, feeling or behavioural intention. In contrast to a fact, an opinion cannot be verified independently. Behaviour refers to the completed action of the survey's subject. For example, John has a GPA of 4.9/5.0 and he never skips any classes. It is a fact that John has a GPA of 4.9/5.0. It is an opinion that John is a good student because he has a GPA of 4.9/5.0. Another opinion is John is a good student because he never skips classes. However, never skipping classes is a behaviour.

Surveys can be conducted through electronic surveys, mail surveys, telephone interviews and face-to-face interviews. The type of survey depends on the survey topic. The success of a survey research depends on sampling, measurement and overall survey design. There are many different types of sampling techniques: simple random sampling, cluster sampling, stratified sampling, systematic sampling, multistage sampling, convenience sampling and purposive sampling. For more details on sampling techniques, refer to Levy and Lemeshow (2009) and VanderStroep and Johnson (2010). Two types of measurement errors are common in survey research: observation errors and non-observation errors (Check and Schutt, 2012; Fowler, 2014).

Observation errors result from the way questions are posed, the character-istics of the survey subjects and the interviewers. Non-observation errors result from unrepresentative samples, sampling errors and non-response.

To minimise observation errors, the survey instrument should be care-fully considered. The survey instrument refers to the topic, the instructions, the layout and the order of the questions. The questions should be designed such that they address the research. Assurances of anonymity and confi-dentiality in the instructions are important to eliminate non-responses or response biases in the survey. The layout of the survey questions should be neat, clear, clean and spacious. The questions should not be vague and each question should have a frame of reference on how the survey subject should answer the question. The wording of the questions should avoid negative words and double negatives. Avoid double-barreled questions, which are questions that involve more than one issue but allow only a single answer (Babbie and Benacquisto, 2009) and do not use words that may lead to response bias.

The questions can be both closed and open-ended. Closed-ended ques-tions require an explicit response and it is common to use a rating scale such as Likert scale (Likert, 1932) for the response. When using such a scale, consider provisions for a neutral response (from fence-sitters) and a nil response (from floaters). Open-ended questions require the survey sub-ject to provide their own response. Usually open-ended questions are used when little is known of the topic. Responses to open-ended question are time-consuming and difficult to analyse.

The order of questions in a survey is also important and should follow a logical order. For example the questions can be sorted into thematic catego-ries and then separated into logical sections in the questionnaire. Usually it is a good idea to pretest the questionnaire with a small group before admin-istering the questionnaire to the whole sample group. For guidelines on min-imising non-observation errors, see the discussion of sampling method and sampling bias in the section 'Experimental research' earlier in this chapter.

REFERENCES

Babbie, E. R., & Benacquisto, L. (2009). *Fundamentals of Social Research*. London: Wadsworth Cengage Learning.

Borrego, M., Douglas, E. P., & Amelink, C. T. (2009). Quantitative, qualitative, and mixed research methods in engineering education. *Journal of Engineering Education* 98(1): 53–66.

Check, J., & Schutt, R. K. (2012). *Research Methods in Education*. Thousand Oaks, CA: Sage.

Cochran, W. G. (1963). *Sampling Techniques* (2nd ed.). New York: John Wiley & Sons.

Cook, T. D., & Campbell, D. T. (1979). *Quasi-Experimentation: Design and Analysis for Field Settings*. Chicago, IL: Rand McNally.

Dane, F. C. (1990). *Research Methods*. Pacific Grove, CA: Brooks/Cole.

Edmonds, W. A., & Kennedy, T. D. (2013). *An Applied Reference Guide to Research Designs: Quantitative, Qualitative, and Mixed Methods.* Thousand Oaks, CA: Sage.

Fowler Jr., F. J. (2014). *Survey Research Methods* (5th ed.). Thousand Oaks, CA: Sage.

Hitchcock, G., & Hughes, D. (1995). *Research and the Teacher: A Qualitative Introduction to School-based Research* (2nd ed.). London: Routledge.

Krejcie, R. V., & Morgan, D. W. (1970). Determining sample size for research activities. *Educational and Psychological Measurement* 30: 607–610.

Levy, P. S., & Lemeshow, S. (2009). *Sampling of Populations: Methods and Applications* (4th ed.). New York: John Wiley & Sons.

Likert, R. (1932). A technique for the measurement of attitudes. *Archives of Psychology* 140: 106–116.

Neuman, W. L. (1994). *Social Research Methods* (2nd ed.). Boston, MA: Allyn & Bacon.

VanderStroep, S. W., & Johnson, D. D. (2010). *Research Methods for Everyday Life: Blending Qualitative and Quantitative Approaches.* New York: John Wiley & Sons.

Chapter 8

Keeping research records

Ideas are the most fragile things in the world, and if you do not write
them down, they will be lost forever.

Phil Cooke
One Big Thing: Discovering What You Were Born to Do, 2012

A research project has a longer duration than any course assignment that
you would have done as it involves much information gathering and research
activity. It is good practice as well as highly recommended that you keep
a record of all your research activities. This record is commonly known as
a research log, research notebook or research diary. A *research notebook*
is usually used by full-time research workers to record their daily activities
that are verified by other research workers, usually their supervisors. A
research diary is associated with the diary method of research that involves
participants keeping a record of events or personal accounts and is more
common in the social sciences (Sheble and Wildemuth, 2009). A *research
log* is a chronological record of all research activities. For a short-term
research project, a *research log* is perhaps more appropriate. We will focus
mainly on research logs in this chapter.

PURPOSE OF KEEPING A RESEARCH LOG

There are many good reasons for keeping a research log:

- It provides a record of all the ideas or work that you have (or have not
  done) which you can refer to from time to time.
- It provides a record of all the information that you will need when you
  write up the project.
- It tracks the research process so that you will be able to avoid repeating
  an experiment or a mistake.
- It enables you to keep track of the progress of the research project, specifically, what has been accomplished and what else needs to be done.
- It provides a basis for you to reflect on or review your project. This is
  especially useful when you seem to have come to a dead end.

The research log thus serves as a 'beacon of light' when you become 'lost' in the research process and are unsure of the way forward. At such times, you should refer to your research log to enable you to answer questions such as 'Is solving the current problem crucial to the successful outcome of the project?', 'What have I tried doing so far that is not working?', 'What else have I not tried?' or 'Is there an alternative method?'

Although it is not common to arrive at a negative outcome with engineering research, it can happen, especially with an undergraduate project that is constrained by time. Should the outcome of your research project be negative, your research log will be important as you can still produce a report documenting the work you have done that produced the negative outcome and recommending what could be done in the future to achieve a successful outcome. Thomas Edison, who encountered numerous failures before achieving success in his invention of the light bulb said, 'I have not failed. I've just found 10,000 ways that won't work.'

The research log is especially useful when you are at the stage of writing up the project as all the important materials you need are recorded in one place. Many students find that a research log reduces the amount of time spent on writing the report or thesis as some parts from the research log can be directly transferred to the report or thesis.

CONTENTS IN A RESEARCH LOG

The research log typically contains the following:

- Notes from meetings with your supervisor
- Records of references that you have collected
- Summaries of literature that you have read with your comments on it
- Problems that you have encountered
- Questions that have arisen
- Ideas or inspirations that you may have
- Sketches of experiment set-ups
- Experimental procedures
- Notes on methodology
- Records of experiments and results
- Observations and conclusions
- To-do lists or action plans

USING A RESEARCH LOG

Conducting research is very different from taking a course. A research project is unstructured and it is likely that you will be thinking about it constantly, so it is useful to have the research log within arm's reach. Keeping a research log will allow you to jot down any idea or inspiration that you may have so that you can explore it later. You should always record your thoughts in your research log at the earliest opportunity so that any details important to the project are noted down. Seemingly trivial entries may turn out to be more useful than originally thought.

WRITING A RESEARCH LOG

There is no fixed format on how to write a research log. As you are usually the one who will use the research log, you can adopt any format that appeals to you. Some simple rules for writing a research log are as follows:

- Write your *name and contact details on the front cover* of the research log so that if you accidentally lose it, it may be possible for someone who finds it to return to you. You may want to preface your name and contact details with 'If found, please contact the following: ...'
- Write the *title of your project, objective and scope on the first page* of your research log. You should regularly read this page to stay on track.
- Include the *date and time for each entry*. In some situations, it may be helpful to be able to recall the date and time when an entry was made.
- If an entry is long, a *subject title* will be useful to summarise the content of the entry.
- Have a *consistent format for each type of entry*. For example, when recording an experiment, start by stating the objective of the experiment and then go on to describe the experimental set-up, the procedures, the results and the conclusion.
- Crossed out *changes* made to an entry instead of using white-out as this will allow you to recall what was previously entered. You should never tear out and throw away any pages from a research log.
- Draw a horizontal line as a *separator* after each entry to separate it from the next entry.

RECORDING LITERATURE SEARCHES

Conducting literature searches is an important part of research. Literature searches should be conducted periodically to ensure that you keep abreast with the latest research. Literature searches should be recorded in the research log. When recording electronic literature searches, make sure that you also record the following details in your literature search:

- Date and time that you conduct the electronic literature search.
- The search engine or database that you use.

- The keywords (including Boolean operators) that were used in the search.
- The results of the search. Such information can be copied into an electronic research log or printed and then pasted into a hardcopy research log.

HARDCOPY VERSUS ELECTRONIC COPY

An electronic research log enables you to 'cut and paste' information from electronic sources into the research log and later from the research log into the final report. However, an electronic copy may not offer the same convenience and accessibility as a hardcopy for recording an experiment. A hardcopy research log allows you to scribble information any time and any where. It is easier to sketch out one's thoughts and ideas in a hardcopy research log than in an electronic research log. It is, therefore, highly recommended that a traditional notebook be used as a research log. Do NOT use a binder with loose sheets of paper as a research log since loose sheets may easily get lost. Messy sheets from a binder are more likely to be thrown out. However, they may contain important information that may not be apparent when they are discarded. Many companies produce dedicated research notebooks for serious researchers. Some of these notebooks even have waterproof pages that can withstand wet conditions in the laboratory or in the field.

THINGS TO AVOID IN A RESEARCH LOG

Do not use a pencil for making entries in a hardcopy research log. Avoid using a pen that could smudge an entry in a hardcopy research log. Try to use the same pen for same day entry to avoid giving the impression that an entry has been altered at a later date. Avoid deleting entries in a research log by using white-out; instead cross out the entries so that you can recover them when you have second thoughts. Avoid the urge to rip out untidy papers from a research log. A research log is expected to be untidy as it is where you write your thoughts, ideas, experiments, calculations, meetings and so forth. Similarly, entries in an electronic research log should not be deleted or altered. It is easy to alter an electronic entry but once altered, that entry will be lost. Instead you should follow the same rule for an entry in a hardcopy research log, that is, strike a line through the entry (use strikethrough font) so the original entry is visible. When there is a need to prove authenticity and originality of research, an entry in a hardcopy research log is considered more credible than an entry in an electronic research log unless this is accompanied by a time stamp and it can be proven to be unalterable.

RULES ON KEEPING A RESEARCH LOG

Different universities, research institutions and organisations may have rules on keeping and maintaining a research log. You should check with your supervisor on the rules that apply in your institution.

A research log is a very important document of your research and you should guard it zealously against loss or theft as it would mean losing practically all the valuable information you have gathered in your research. For full-time researchers, a witnessed and signed research log can serve as proof of ownership of original ideas or discoveries and can be produced as evidence of such claims in court. For the same reason, if you choose to keep your research log in electronic form you will need to take precautions to have backup copies so that it will not be accidentally deleted or lost due to failure of the storage device. Furthermore, if you keep an electronic research log, it is important that your research log remains private until your research is published to maintain the originality and integrity of your research (especially if you are a PhD student).

You may also want to read *Writing the Laboratory Notebook* by Kanare (1985) for a more comprehensive understanding of the research log.

INTERESTING FACTS

Michael Faraday, who made several important discoveries including the induction motor, kept meticulous notes of his experiments and detailed sketches of his apparatus in what are now known as Faraday's notebooks (Tweney, 1991). It was surmised that Faraday was mistrustful of his own memory and kept records so that he would not forget experiments that he had previously conducted (Hare, 1974; Williams, 1964).

Leonardo da Vinci, well-known for his painting titled Mona Lisa, was also a sculptor, architect, musician, scientist, mathematician, engineer, inventor, anatomist, geologist, cartographer, botanist and writer. Leonardo made meticulous notes and drawings. He guarded his writings zealously by adopting a mirror-cursive writing style that is difficult for others to read. Leonardo's notes are written in a form that suggests that he intended them for publication (Arasse, 1997).

There are several good books that illuminate the creative thought processes. *Notebooks of the Mind: Explorations of Thinking* by Johnson-Steiner (1996) attempts to understand the creative processes of artists, philosophers, writers and scientists arising from accumulations of fragmented ideas. *Where Good Ideas Come From: The Natural History of Innovation* by Johnson (2010) provides a new understanding on innovation from the past to the present and identifies the key traits behind creative breakthroughs.

REFERENCES

Arasse, D. (1997). *Leonardo da Vinci*. Old Saybrook, CT: Konecky & Konecky.

Hare, E. H. (1974). Michael Faraday's loss of memory. *Proceedings of the Royal Society of Medicine* 61: 617–618.

Johnson, S. (2010). *Where Good Ideas Come From: The Natural History of Innovation*. New York: Penguin Group.

John-Steiner, V. (1996). *Notebooks of the Mind: Explorations of Thinking*. New York: Oxford University Press.

Kanare, H. M. (1985). *Writing the Laboratory Notebook*. Washington, DC: American Chemical Society.

Sheble, L., & Wildemuth, B. (2009). Research diaries. In B. Wildemuth (Ed.). *Applications of Social Research Methods to Questions in Information and Library Science* (pp. 211–221). Santa Barbara, CA: Libraries Unlimited.

Tweney, R. D. (1991). Faraday's notebooks: the active organization of creative science. *Physics Education* 26: 301–306.

Williams, L. P. (1965). *Michael Faraday: A Biography*. New York: Basic Books.

Part II

Writing

Chapter 9

Starting to write

Begin with the end in mind.

Steven R. Covey
The 7 Habits of Highly Effective People, 1989

Writing is an essential part of a research project. It is through your writing that you communicate the findings from your research. The final report in an undergraduate programme is usually called the 'Final Year Project Report' (FYP report) and that in a postgraduate programme is referred to as the thesis or dissertation. In this book, we will use the term *report* to signify the end report of an undergraduate research project and *thesis* to refer to the end report of a postgraduate research project. In a postgraduate programme, there may also be interim reports and academic papers to write before the thesis is submitted. The report or thesis, however, is probably the longest piece of writing that you will be required to do in your university studies. Whether you are writing a report or thesis, it is the basis on which the value of your work is judged as well as the basis on which your grade or degree is awarded.

Despite the importance of the report or thesis, many students do not give it the time and the attention it deserves. Students tend to leave their report or thesis to the 11th hour, resulting in insufficient time given to writing a well-planned and coherent report or thesis. It is sad to see students who have worked very diligently on a project to be let down at the end by a poorly written report or thesis. Though no amount of good writing will make up for poor content, good content can be undermined by poor writing.

STARTING TO WRITE

Students tend to delay writing until a deadline looms on the horizon. That deadline may seem very distant at the start of a project. Students, especially engineering students, often find writing difficult. Writing is perceived as a chore to be avoided especially at the start of the project when there is not much to write about. Although the writing at the beginning may only be the fragments of ideas and the rudiments of your initial knowledge of the project, it is still important to write. Remember that 'A journey of a thousand miles begins with a single step' (Lao-tzu, Chinese philosopher, 604 BC–531 BC). Once you start writing, you will find that the writing process gets easier with time. Keep in mind that good writing involves rewriting. The process of writing a good final report involves drafting, revising and editing (see *Chapter 20—Revising and editing*). A well-written report or thesis is one that has gone through several drafts before submission.

Writing regularly is good practice as writing and thinking go together. Through writing, you can clarify your ideas and thoughts, identify problems as well as refine the planning of your project to ensure greater progress. Writing in a research log (described in *Chapter 8—Keeping research records*) is a good way to start writing your report or thesis.

WRITING AT THE BEGINNING OF A PROJECT

At the beginning of a project, you can start by writing the title of the project, a brief description of the project, the objective(s) and the scope of the project the Such information may change over the course of the project, which is quite usual in research and if it does, you should rewrite it. You can also start creating an outline of the final report or thesis that you will need to submit. It is helpful to refer to past reports or theses in your discipline to get an idea of how they are structured. Generally, an engineering report or thesis has the following main chapters:

- Introduction
- Literature Review
- Materials and Methods
- Results and Discussion
- Conclusion
- References

The chapter titles can form the basic outline of your report or thesis. The Materials and Methods and Results and Discussions chapter titles may be replaced with more specific titles depending on the nature of your project. Under each chapter, you can have subsections with their own headings. The headings will come to mind as you make progress in your project.

As the report or thesis may be lengthy, it is a good idea to create one document file for each chapter before starting to write. Create the document files by using the chapter title to name the file, for example Introduction. doc or Introduction.docx in Microsoft Office Word. You can then start writing the background, objective(s) and scope of your research project in the Introduction document.

MANAGING YOUR WRITING

Writing the draft report or thesis is different from writing in the research log. The draft report or thesis is written over longer time intervals whereas the research log records your research activities and are made during or immediately after an activity. Writing the draft report or thesis consists mainly of transferring your entries in the research log into the draft report or thesis at regular intervals. While the research log contains all manners of information relating to the research project, not all entries will be transferred into the draft report or thesis. Only the distilled ideas, experimental details and results will be transferred. It is a good practice to review your research log periodically, say fortnightly or monthly, and transfer relevant entries into your draft report or thesis. Unlike the research log, your writing in the draft report or thesis should be in full sentences and paragraphs. These should contain developed ideas and arguments, synthesised and consolidated from your entries in the research log.

You can also adopt the process above for writing academic papers and interim reports. Academic papers and interim reports tend to be much shorter in length than the final report. For academic papers, the focus is usually narrower and you probably can keep all the contents within one document file. Interim reports may also be abstracted directly from the draft report or thesis.

AVOIDING PROCRASTINATION

Procrastination is putting off to a future date what you need to do today. It is easy to put off writing when you do not need to hand in the report or thesis until sometime in the future, and that future time can be many months away. However, failure to start writing as early as possible will translate to an emergency at a later date. The consequences are a poorly written report or thesis and a poor grade even though you have conducted a brilliant piece of research work.

To avoid procrastination, planning and time management are essential. Just as for the courses that you are attending or have attended, your project deserves to be allocated dedicated times in your timetable. If you have to divide your time between research and course work, the minimum amount of time to be allocated to a project should be based on the academic weighting of your research project. If the weighting of the research project is equivalent to three courses, you should set aside time in your timetable that is at least equivalent to class time for three courses. Furthermore, it is advisable to include in the time allocated for your project time that you would normally set aside for review and assignments in your other courses. Once you have allocated project times in your timetable, you should adhere to it strictly. Within the allocated project times, set aside specific times in your timetable fortnightly or monthly for writing of the draft report or thesis and be sure to stick to the plan.

In addition to writing regularly, you should also get regular feedback on your writing so that your writing skills improve over time. The most appropriate person to give you feedback on your writing is your supervisor. Should your supervisor be too busy to read your writing at frequent intervals, you could still bring along what you have written to your regular meetings with your supervisor to discuss the contents. By sharing what you have written with your supervisor, you will be able to gauge if you have made progress in the project. Thus writing promptly and consistently will also help to keep you on track besides reducing your stress as the deadline for submitting your final report or thesis approaches.

INTERESTING FACTS

Ernest Hemingway, winner of the Nobel Prize in Literature 1954 for his short novel *Old Man and the Sea*, had a routine of waking up at 7 AM every day and writing between 500 and 1,000 words a day without fail.

Stephen King, a prolific author of horror novels such as *Carrie, The Shinning*, and *Pet Sematary*, mentioned in his book, *On Writing: A Memoir of the Craft*, that he writes 10 pages a day without fail, even on holidays.

Alfred Russel Wallace (1823–1913), a British naturalist, explorer, geographer, anthropologist and biologist who is best known for independently conceiving the theory of evolution through natural selection, published more than 700 papers of which 191 appeared in *Nature*. Considering the period over which these publications appeared, Wallace was publishing at the astonishing average rate of one publication per month before the computer was even invented!

Chapter 10

Grammar, punctuation and word usage guide

> The flesh of prose gets its shape and strength from the bones of grammar.
>
> Constance Hale
> *Sin and Syntax: How to Craft Wicked Good Prose, 2013*

This chapter offers guidance in language areas where students commonly make mistakes when writing. It provides general rules on usage as well as advice on how to avoid common errors in grammar, sentence construction, punctuation and word usage.

COMMON GRAMMAR PROBLEMS

Tenses: Present, past, future

The tenses that you will need to use in writing your report or thesis are primarily the simple past, the simple present and the present perfect. Writing Guidelines 10.1 provides an overview of the key rules for these tenses, which are typically used in technical and scientific documents (Wallwork, 2013, pp. 49–58).

Writing Guidelines 10.1 Tense usage in technical and scientific documents

Tense	Used for	Example sentences
Simple present	An action or situation which happens regularly, repeatedly, or is always/generally true. This tense is typically used to state established scientific facts and findings, theories, definitions and so on.	Air *is* 79% nitrogen. Figure 1 *shows* ... This section *presents* ... This type of vinegar *contains* about 3% acid.
Present perfect	An action, event, or scenario that happened at an unspecified time in the past, or that began in the past and is still current today. This tense is used to announce a new finding or an advance in a particular discipline.	Researchers *have developed* a new system for converting ocean wave motion into energy. The sea level *has changed* throughout the Earth's history and will continue to do so.
Simple past	An action or event that is clearly past, or that has a clear time reference (e.g. *2013, three months ago*, etc.)	Marie Curie *conducted* pioneering research on radioactivity. Watson (2000) first *used* this procedure more than a decade ago.
Simple future *(will)*	Making predictions, announcing future work/research, talking about later parts of the document	Demand *will outweigh* supply and prices will rise. Future work *will involve* investigating the causes for these defects. The final chapter *will discuss* the implications of these findings.

The following are suggestions for deciding on the tenses to use in the different sections or chapters of your document:

Chapter/section	Tenses	Use in report/thesis
Introduction	Simple present; present perfect; simple past	• Use the **simple present** in the **Background** when describing an existing situation or universal truth or condition. (e.g. *Clean water is a basic human need.*) The **present perfect** is used when describing a state which began in the past and continues to the present. (e.g. *Its discovery, transport and systematic renewal* **have** always **been** *crucial to human societies.*) • When describing the *aims* of your project, use the **simple present** if the project is ongoing. (e.g. *The aim of this study is to ...*) Use the **simple past** if your report/paper describes a completed project. (e.g. *The aim of this project was to ...*) • The **simple present** is used when outlining the structure of the paper. (e.g. *Section 1* **demonstrates** *that the LSNDPD is NP-hard.... Section 2* **elaborates** *liner shipping network design problem ...*)

Writing Guidelines 10.1 (Continued) Tense usage in technical and scientific documents

Chapter/section	Tenses	Use in report/thesis
		(Continued)
Literature Review	Simple present; simple past; present perfect	• The *simple present* is used when the information cited is generally accepted as a scientific fact (e.g. The factors that control the concentration of aluminium in seawater are poorly known [3]). • When you believe that the findings reported are restricted to the specific study you are citing, use the *simple past*. (e.g. *Hamill (1998) found that ...*) When you believe that the findings reported are not restricted to the specific study or have relevance to your own study, use the *present perfect* (e.g. *Bergh and Magnusson (1987), Verhey (1983) and Hamill (1999) have experimentally studied propeller wash.*) • The *present perfect* is used in citations where the focus is on the research area of several authors. (e.g. *Several researchers have studied the relationship ... [5, 7, 10].*) The present perfect is also used in general statements that describe the level of research activity in an area. (e.g. *Much research has been done on ...*)
Materials and Methods	Simple past; simple present	When describing the procedure in the **Methods** section, use the *simple present* if the procedure or method you used has been established by other researchers. (e.g. *Our methodology consists of ...*) Use the *simple past* to state the objectives of your experiments, the equipment you used, the steps followed, how other methods were adapted and so on. (e.g. *The aim of our procedure was .... We adapted the software by ...*)
Results and Discussion	Simple present; simple past	• Use the *simple present* to state what a figure or table shows, highlights, describes and so on (e.g. *Figure 4.1 shows that ...*). • Use the *simple present* to discuss your results, findings, and to state their implications, typically after verbs such as show, explain, highlight, indicate, reveal and so on (e.g. *This finding indicates the need for ...*). • Use the *simple past* to talk about actions you performed—what you found, discovered, noticed, observed and so on (e.g. *We observed/It was observed that ...*).
Conclusion	Simple past; future tense with 'will'	• In the **Conclusion** section, will is used to outline future work (e.g. *Future work will investigate/ focus on ....*).

Subject–verb agreement

Subject-verb agreement errors are very common in writing. To avoid such errors, start by determining if the subject of the sentence is singular or plural. Then make sure the verb *agrees in number* with its subject, that is, singular verbs should be used with singular subjects, and plural verbs should be used with plural subjects. For example:

This study investigates the development of a scour hole with non-cohesive sediments.

The results indicate that scour depth increases with an increase in propeller speed.

This may sound straightforward; however, it is easy to make mistakes when the subject is more complex than those in the examples above. For example:

The integration of the telecommunication systems *were* necessary.

In this example, the plural verb 'were' is incorrectly used. This is because the verb should agree with the head noun 'integration', which is singular. The correct verb form should be 'was' as in:

The integration of the telecommunication system *was* necessary.

Two other areas where subject-verb agreement errors are made are as follows:

* Indefinite pronouns: either, neither, each

   The indefinite pronouns *either, neither, and each* are always used with singular verbs:

   Incorrect: Neither of the models *are* applicable.

   Correct: Neither of the models *is* applicable

   Incorrect: Each of the models *have* flaws.

   Correct: Each of the models *has* flaws.

* Compound subjects

   When *or, either ... or*, or *neither ... nor* are used to create a compound subject, the verb should agree with the last item in the subject:

   Neither the circuit nor *the transistors were* modified.

   Note: If the compound subject comprises a singular noun and a plural noun (circuit, transistors), the plural noun should come after the singular noun so that the sentence sounds less awkward.

   Awkward: Neither the transistors nor *the circuit was* modified.

Modal verbs

When reporting your own study, it is advisable to use a cautious tone. This is achieved by using modal verbs. Common modal verbs used in reporting research are *will, would, should, may, might, can, and could*. These modals indicate different degrees of certainty. Choosing the appropriate modal verb can be difficult as the meaning of some of these words differ

MODAL VERBS: Degrees of Certainty

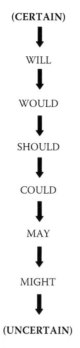

(CERTAIN)

WILL

WOULD

SHOULD

COULD

MAY

MIGHT

(UNCERTAIN)

Figure 10.1 Modal verbs: degree of certainty.

only slightly from one another. The following chart lists the modal verbs in order of their degree of certainty (Figure 10.1).

Examples:
 Will (very certain about the future):
  The findings *will* supplement those from previous studies.
 Would (certain about the future, assuming certain conditions):
  The purpose of this study was to determine if ... *would* ... improve ...
 Should (reasonable expectation about the future):
  This modification *should* simplify the analysis procedure.
 Could (some uncertainty about the future):
  Results of this study *could* have considerable impact on estimat-
   ing the rate of global warming.
 May (more uncertainty about the future):
  Both of the factors studied here *may* be of importance in explain-
   ing the causes of aviation disasters.
 Might (uncertain about the future):
  It *might* be possible to repair the engine (but is highly unlikely as
   the damage is extensive).

Pronoun reference

Pronouns are used as substitutes for nouns; however, when used incorrectly, pronouns can obscure the meaning of a sentence. Common pronoun errors include

- Unclear pronoun reference
  A pronoun must refer to a specific noun (the antecedent). Ambiguous pronoun reference confuses readers and obscures the intended meaning.
  Problem: Writers should set aside time for revising their documents to make sure they are perfect. *(Unclear antecedent: who or what are perfect—writers, documents?)*
  Revision: Writers should set aside time for revising their documents to make sure *these documents* are perfect.
- Vague subject pronoun
  Pronouns such *as it, this, and there* are often used as subjects in sentences. Make sure when using a pronoun as subject that its antecedent is clear.
  Problem:
  The reaction between nitric oxide and ozone results in the production of nitrogen dioxide (NO_2) and molecular oxygen. *It* plays an important role in the chemistry of the ozone layer that surrounds the earth and protects us from the sun's harmful ultraviolet radiation. *(What does 'It' refer to? Molecular oxygen? Production of nitrogen dioxide (NO_2) and molecular oxygen? The reaction between nitric oxide and ozone? The entire preceding sentence?)*
  Revision:
  The reaction between nitric oxide and ozone results in the production of nitrogen dioxide (NO_2) and molecular oxygen. *This process* plays an important role in the chemistry of the ozone layer that surrounds the earth and protects us from the sun's harmful ultraviolet radiation.
- Pronoun agreement error
  A pronoun must agree with its antecedent in gender and number. Often, pronouns do not agree with their antecedents in number. For example:
  Problem: Apple has launched their new iPhone 6. *('Apple' is a singular noun while 'their' is a plural pronoun.)*
  Revision: Apple has launched its new iPhone 6.

Passive and active voice

In English, all sentences are formed in either the active or passive voice:

Examples:
  Active: We analysed the data.
  Passive: The data were analysed (by us).

In an active sentence, the person or object responsible for the action (the actor) in the sentence is mentioned first. However, in a passive sentence, it is the person or object receiving the action that is mentioned first, and the actor is added at the end, introduced with the preposition 'by'. In passive sentences, the actor is often not mentioned (*e.g. The data were analysed*). The passive form of the verb is indicated by a form of the verb 'to be'. In the preceding examples of active and passive sentences, 'were analysed' is in the passive voice while 'analysed' is in the active voice.

When should you choose the passive voice and when the active? To answer this question, you need to know the difference between them.

The passive voice emphasises the person or object receiving the action by placing the recipient of the action near or at the beginning of the sentence in the topic position *(e.g. The data were analysed)*. In contrast, the active voice emphasises the person or object performing the action *(e.g. We analysed the data)* by placing the actor first.

Reasons for using the passive voice

1. To emphasise the person or object acted on: the recipient is the main topic.
   For instance, if we were writing a paragraph where the focus is on wave energy, we would start with the following passive-voice sentence:
   Wave energy is generated by the wind as it blows across the sea surface.
   However, if the focus of the paragraph is on wind and its relationship with wave energy, we would write an active-voice sentence:
   Wind generates wave energy when it blows across the sea surface.
2. When the actor is unknown or irrelevant.
   The building was demolished in the 1800s. (We do not know by whom.)
   Up to 90% of the energy in conventional light bulbs is wasted in the form of heat. (The actor is irrelevant to the discussion.)
3. When writing in a genre where the convention is to use the passive voice.

Passive voice is often preferred in scientific research papers and reports, especially in the Materials and Methods section, as it places the emphasis on the experiment or procedure rather than on the researcher. An emphasis on the researcher is not desired as the general topic is the research materials and procedure.

COMMON SENTENCE PROBLEMS

Sentences provide us with the framework for expressing our ideas clearly. A sentence which is well constructed is complete and correctly punctuated. A complete sentence always contains a subject, a verb, expresses a complete idea and makes sense standing alone. The following are common sentence structure problems:

- Fragments
- Comma splices and run-on sentences
- Misplaced and dangling modifiers

Fragments

A complete sentence has the following three components:

1. A subject (the actor in the sentence)
2. A verb
3. A complete thought (that is, the sentence can stand alone and make sense)

If any of these is missing from the sentence, it is a fragment.

> Fragment: The types of applications range from games to office suites. *Enabling the owner to perform a wide array of tasks.*
> The second sentence is a fragment, because it has neither a subject nor a verb. (A participle such as 'enabling' is verbal or 'part-verb', not a verb.) The fragment can be made into a complete sentence by adding a subject and a verb as in the revision below.
> Revised sentence: The types of applications range from games to office suites. *These applications enable the owner to perform a wide array of tasks.*
> Another way to correct the fragment is to combine it with the preceding sentence by replacing the full stop with a comma and thus turning the fragment into a participial phrase.
> Revised sentence: The types of applications range from games to office suites, *enabling the owner to perform a wide array of tasks.*
> Be careful not to separate subordinate (dependent) clauses from the main clause. Subordinate clauses begin with subordinating conjunctions such as *whereas, while, because, although, since, as, that, and until.*
> Note: A *dependent clause (or a subordinate clause)* is a clause that augments an *independent clause* (a clause that is a complete sentence) with additional information A dependent clause cannot stand alone as a sentence.
> Fragment: The university wants to introduce more online courses. *Although a number of faculty are sceptical.*
> Revised sentence: The university wants to introduce more online courses *although* a number of faculty are sceptical.

Run-on sentences and comma splices

A run-on sentence is made up of two (or more) complete sentences jammed together without any punctuation and without any conjunction.

Run-on sentence: Glaciers are melting so fast scientists are concerned about rising sea levels.
Revised sentence: Glaciers are melting so fast *that* scientists are concerned about rising sea levels.

In the revised sentence, the problem has been corrected by adding the word *that*, which turns the second independent clause (scientists are concerned about rising sea levels) into a dependent, or subordinate, clause.

Comma splices

In the typical *comma splice* sentence, two sentences are joined by a comma without an intervening coordinating conjunction (*and*, *or*, *nor*, *but*, *yet*).

Comma splice:
Books, generally, do not have the most up-to-date information, it is a good idea then to look for articles in specialised journals.

There are three ways of correcting the comma splice sentence above:

1. By replacing the comma with a period after *information* and starting a new sentence:
   Books, generally, do not have the most up-to-date information. It is a good idea then to look for articles in specialised journals.
2. By replacing the comma with a semicolon:
   ... information; it is a good idea ...
3. By making one of the independent clauses subordinate to the other, so that it cannot stand by itself:
   As books, generally, do not have the most up-to-date information, it is a good idea then to look for articles in specialised journals.

Misplaced and dangling modifiers

Modifiers describe, clarify or give more detail about another word in the sentence. However, if they are placed in the wrong location in the sentence, they give rise to sentences which are illogical, confusing or absurd.

Misplaced modifiers

A misplaced modifier is a word or phrase that refers to the wrong word or phrase in a sentence. The error is caused by locating the modifier in the wrong place within the sentence.
Example: The defective equipment was returned to the company which manufactured it by courier.

Does the modifier 'by courier' in the preceding sentence refer to the manner in which the equipment was shipped or manufactured? Because 'by courier' is placed far from the word it modifies ('was returned'), the sentence is ambiguous.

Revised Sentence: The defective equipment was returned by courier to the company which manufactured it.

To avoid misplaced modifiers and confusing your readers, place modifiers next to the words or phrases they modify.

Limiting modifiers such as *only, almost, nearly, just, hardly, and even* are commonly misplaced.

To avoid ambiguity, place the limiting modifier in front of the word it modifies so that there is no possibility for misinterpretation. The sentences below convey different meanings as the modifier *only* is placed before a different word in each of the sentences:

- *Only* this study investigates the fluctuations of wind power.
- This study *only* investigates the fluctuations of wind power.
- This study investigates *only* the fluctuations of wind power.
- This study investigates the *only* fluctuations of wind power.

Dangling modifiers

A modifier is said to be *dangling* if the word that the writer intends to modify is unclear or missing. The result is an illogical statement as in the following example:

After completing the project, the results were published.

The introductory phrase 'After completing' expresses an action but the actor cannot be 'the results'. Since the actor of the action mentioned in the introductory phrase has not been stated, the phrase is said to be a dangling modifier.

The dangling modifier above can be revised in one of the following ways:

1. Make the logical actor the subject of the main or independent clause ('After completing the project, *the researcher* published the results').
2. Change the phrase that is dangling into a complete clause by naming the actor in that clause ('After *the researcher* completed the project').

Dangling modifiers tend to occur when writing is wordy and the passive voice is used inappropriately. To avoid dangling modifiers, make your writing concise and use the passive voice with care.

COMMON PUNCTUATION PROBLEMS

This section provides guidelines for avoiding common punctuation errors in writing. It covers:

- Commas
- Semicolons and colons
- The apostrophe

Commas (,)

A comma indicates a pause. Commas are generally used as follows:

- Use a comma after each item in a series (but generally not before the final item preceded by 'and')
  The experiments were carried out in a tank *1.8 m wide, 4 m long and 1 m deep.*
- Use commas to separate a dependent clause from the main clause in a sentence
  The evolution of a typical scouring profile along the longitudinal direction, *which is illustrated in Fig. 2,* may be divided into four stages.
- Use a comma after an introductory phrase or dependent clause
  To minimise propeller-induced scour and better protect port facilities, an improved understanding of scour due to propeller wash is imperative.
  Although the Kyoto Protocol was signed in 2007, it has had limited impact on reducing global warming.
- Use commas to separate information that is not essential to the meaning of the sentence
  This rapid growth in human population, with the accompanying urban migration and industrialisation, has impacted water ecosystems around the world.

Semicolons (;)

Semicolons are used to connect independent sentences to show a close relationship between these sentences. Semicolons are generally used as follows:

- Use a semicolon between two closely related sentences.
  The Aral Sea, once the fourth largest lake in the world, has been evaporating and shrinking since the 1960s; the lake's area is now 25% of its original size and holds just 10% of its original volume of water (Visible Earth, 2015).

In the example sentence above, the statement on each side of the semi-colon is able to stand alone as a sentence.
- Use semicolons to separate list of items in a list when the items in the list are separated by commas
Electric power substations are used for some or all of the following purposes: *connection of generators, transmission or distribution lines, and loads to each other; transformation of power from one voltage level to another; interconnection of alternate sources of power;* and detection of faults, monitoring and recording of information, power measurement, and remote communication (Electric power substation, n.d.).

Colon (:)

A colon indicates that something will follow: a list, a series, or an elaboration.

- Use a colon before a numbered or bulleted list

The process may be divided into four stages:

1. Initial stage
2. Developing stage
3. Stabilisation stage
4. Asymptotic stage

- Use a colon before a series
In physics, matter exists in four states: solid, liquid, gas or plasma.
Hurricane size is expressed in three ways: the strength of the maximum winds, the diameter of the hurricane-force winds, the diameter of the gale-force winds, and the overall size the cyclone circulation (http://www.write.armstrong.edu/handouts/UsingColons.pdf).

Apostrophe (')

Use an apostrophe to indicate the possessive form for nouns.

- Use ['s] with a singular noun: the *camera's features* (the features of one camera)
- Use [s'] with a plural noun: the *cameras' features* (the features in a number of cameras)

Note: An apostrophe is also used in contractions (when two words are merged together) such as *can't, hasn't, isn't* and so on. Contractions are used in informal writing; do not use them in formal writing.

COMMONLY MISUSED AND CONFUSED WORDS AND PHRASES

Word usage

This section identifies common errors in word usage and provides guidelines for avoiding these errors. There are two types of errors:

- Irregular plurals (leading to misspellings)
- Pair of words commonly confused

Irregular plurals

Table 10.1 shows words that have irregular plural forms. They are commonly used in technical and scientific writing and are often misspelt.

Table 10.1 Irregular nouns and their spelling

Singular	Plural
Alga	Algae
Analysis	Analyses
Antenna	Antennae ('antennas' is often used in communication engineering)
Appendix	Appendices
Axis	Axes
Bacterium	Bacteria
Basis	Bases
Criterion	Criteria
Focus	Foci
Formula	Formulae
Hypothesis	Hypotheses
Locus	Loci
Matrix	Matrices
Medium	Media
Nucleus	Nuclei
Phenomenon	Phenomena
Quantum	Quanta
Radius	Radii
Research	Research
Species	Species
Stimulus	Stimuli
Stratum	Strata
Symposium	Symposia
Vertebra	Vertebrae
Vortex	Vortices

Pairs of words commonly confused

This section gives an overview of words and phrases that are often confused or misused. Read through the list periodically to refresh your memory. This will help you avoid mistakes when you are writing.

Absorb/adsorb
> *Absorb* means to take in heat, gas, liquid or other substances.
> *Adsorb* refers to the process whereby atoms or molecules adhere to exposed surfaces, usually of a solid.

Advice/advise
> *Advice* is a noun, *advise* a verb. This distinction is made in British English spelling; in American English spelling, 'advice' is both a noun and verb.

Affect/effect
> *Affect* is a verb meaning 'to influence'. *Effect* can be either a noun meaning 'result' or a verb meaning 'to bring about something', usually with reference to change.

All together/altogether
> *All together* means in a group; *altogether* is an adverb meaning 'entirely'

A lot (not *alot*)
> Write as two separate words: *a lot*

Alternate/alternative
> *Alternate* means 'every other' or 'every second' item in a series (e.g. *Students taking the course are divided into two sections which meet on **alternate** weeks*). *Alternative* refers to a choice between two or more options (e.g. *Students could do an internship as an **alternative** to doing a final year project*).

Beside/besides
> *Beside* is a preposition meaning 'next to' (e.g. *The Student's Centre is **beside** the Health Centre*). *Besides* has two uses: as a preposition with the meaning 'in addition to' (e.g. ***Besides** the lab technician, Jane was the only one in the lab*); as a conjunctive adverb with the meaning of 'moreover' (e.g. *References listed in the report were incomplete; **besides,** many of them were very old*).

Complement/compliment
> As verbs, *complement* means 'to complete' or 'enhance', while *compliment* means 'to praise'.

Continual/continuous
> *Continual* means 'repeated over a period of time'; continuous means 'without interruption'.

Data
> *Data* is the plural of *datum*. Informally, *data* is used as a singular noun, but in formal writing, treat it as a plural (e.g. The data *were* analysed).

Deduce/deduct

To *deduce* is 'to work something out by reasoning'; to *deduct* means 'to subtract or take away from' something.

Dependent/dependant

Dependent is an adjective meaning 'subject to'; *dependant* is a noun meaning someone who is given food, a home or money by another person.

Device/devise

In standard British English spelling, *device* is the noun while *devise* is the verb. For example:

*The company **devised** a new operating system for smartphones.*

*The smartphone has become an essential digital **device** for modern life.*

In American English spelling, *device* is both a noun and a verb.

Different from/different than

Although *different than* is commonly used in informal contexts, *different from* is more appropriate in formal writing (e.g. *The results were **different from** what had been predicted*).

Discreet/discrete

Discreet means 'tactful' or 'prudent' with reference to behaviour; *discrete* means 'separate and distinct' in the context of scientific and technical writing (e.g. *Discrete samples were tested*).

Economic/economical

Both words are adjectives. Use *economic* when referring to the economy and *economical* when referring to savings.

Especially/specially

Especially means 'particularly' (e.g. *Safety is an **especially** important consideration in construction projects*), while *specially* means 'for a special purpose' (e.g. *The device was **specially** designed for the elderly*).

Farther/further

Farther generally refers to distance (e.g. *The spacecraft travelled **farther** than expected*), while *further* indicates extent (e.g. *Further research needs to be conducted*).

Infer/imply

To *infer* means 'to conclude by reasoning' (e.g. *We **infer** from the test results that the cause is a defective gasket*). This word is often confused with *imply* which means 'to suggest' (e.g. *The test results **imply** that the cause is a defective gasket*).

Inflammable/flammable/non-flammable

Despite its *in*-prefix (which generally indicates a negative meaning), *inflammable* is not the opposite of *flammable*: both words describe things that are easily set on fire. The opposite of *flammable* is *non-flammable*. To avoid confusion, it is best not to use *inflammable*.

Follow/following

The correct form is 'as follows' with a colon after and not 'as follow':

Incorrect: The causes of global warming are *as follows:*

Correct: The causes of global warming are *as follow*:

The phrase 'the following' is never used with the plural *-s*

Incorrect: *The followings* are the causes of global warming:

Correct: *The following* are the causes of global warming:

Irregardless/regardless

Irregardless is considered nonstandard English as well as redundant (the prefix 'ir-' and the suffix '-less' both convey the same meaning 'without'). Use *regardless* instead.

Lead/led

These two words are often confused as *lead*, the element (Pb), sounds like *led* (the past form of the verb *to lead*).

Less/fewer

Less is used with singular and uncountable nouns (as in 'less information'). *Fewer* is used with plural countable nouns (as in 'fewer details').

Lie/lay

To *lie* is an intransitive verb, meaning either to 'assume a horizontal position' or 'to tell a lie'; to lay is a transitive verb and means 'to put [something] down'. The changes of tense often cause confusion (see Table 10.2).

Loose/lose

Loose means not restrained or the opposite of tight (e.g. *The lid was loose*) while *lose* means 'to misplace' or 'to cease to possess' (e.g. *The battery was losing its charge*).

Passed/past

Passed is the past form of the verb *pass*.

Examples:

The law that has just been passed should deter distracted driving.

All the students taking the course passed the examination.

Past refers either to time before the present or to movement from one side of a reference point to the other side.

Table 10.2 Meanings of lie versus lay

Present	Past	Past participle
Lie (recline)	Lay	Lain
Lie (deceive; tell a lie)	Lied	Lied
Lay (put [something] down)	Laid	Laid

Examples:
The past year has seen a sharp drop in demand for PCs (denotes time before the present).

The Mekong River flows *past* Vientiane before turning south. (denotes movement from one side of a reference point (Vientiane) to the other side).

Principle/principal
Principle is a noun meaning 'a general truth or law'; *principal* can be used as either a noun or an adjective, meaning 'chief'. For example: The *principal* cause of the error was not practising the *principle* of due diligence.

Rational/rationale
Rational is an adjective meaning 'logical' (e.g. *A **rational** decision was made*); *rationale* is a noun meaning explanation (e.g. *The CEO's memo stated the **rationale** for the decision*).

Than/then
Than is a conjunction linking unequal comparisons (e.g. *A is shorter **than** B*); *then* is an adverb of time or sequence (e.g. *Shorten first A **then** B*).

That/which
That is used to introduce a restrictive clause, while *which* can introduce a clause that is either restrictive or non-restrictive. For example:

Follow the format ***that/which** the company recommends* when you are writing reports (restrictive clause). Note that commas are not used before or after a restrictive clause introduced by either *that* or *which*.

Follow the IEEE format, ***which** the company recommends*, when you are writing reports (non-restrictive clause). Note that commas are used to separate a non-restrictive clause from the rest of the sentence as the information is non-essential to the meaning of the sentence.

INTERESTING FACTS

William Safire, a *New York Times* journalist and speechwriter, compiled a list of grammatical rules for writing entitled 'Fumblerules of Grammar'. These 'rules' are all humorously self-contradictory and were published in his popular column, 'On Language.' The following are the 36 fumblerules as well as an additional 18 rules featured in Safire's book, *Fumblerules: A Lighthearted Guide to Grammar and Good Usage* (published 2002).

1. Remember to never split an infinitive.
2. A preposition is something never to end a sentence with.
3. The passive voice should never be used.

4. Avoid run-on sentences they are hard to read.
5. Don't use no double negatives.
6. Use the semicolon properly, always use it where it is appropriate; and never where it isn't.
7. Reserve the apostrophe for it's proper use and omit it when its not needed.
8. Do not put statements in the negative form.
9. Verbs has to agree with their subjects.
10. No sentence fragments.
11. Proofread carefully to see if you words out.
12. Avoid commas, that are not necessary.
13. If you reread your work, you can find on rereading a great deal of repetition can be avoided by rereading and editing.
14. A writer must not shift your point of view.
15. Eschew dialect, irregardless.
16. And don't start a sentence with a conjunction.
17. Don't overuse exclamation marks!!!
18. Place pronouns as close as possible, especially in long sentences, as of 10 or more words, to their antecedents.
19. Hyphenate between sy-llables and avoid un-necessary hyphens.
20. Write all adverbial forms correct.
21. Don't use contractions in formal writing.
22. Writing carefully, dangling participles must be avoided.
23. It is incumbent on us to avoid archaisms.
24. If any word is improper at the end of a sentence, a linking verb is.
25. Steer clear of incorrect forms of verbs that have snuck in the language.
26. Take the bull by the hand and avoid mixing metaphors.
27. Avoid trendy locutions that sound flaky.
28. Never, ever use repetitive redundancies.
29. Everyone should be careful to use a singular pronoun with singular nouns in their writing.
30. If I've told you once, I've told you a thousand times, resist hyperbole.
31. Also, avoid awkward or affected alliteration.
32. Don't string too many prepositional phrases together unless you are walking through the valley of the shadow of death.
33. Always pick on the correct idiom.
34. "Avoid overuse of 'quotation "marks.""""
35. The adverb always follows the verb.
36. Last but not least, avoid cliches like the plague; They're old hat; seek viable alternatives.

37. Never use a long word when a diminutive one will do.
38. Employ the vernacular.
39. Eschew ampersands & abbreviations, etc.
40. Parenthetical remarks (however relevant) are unnecessary.
41. Contractions aren't necessary.
42. Foreign words and phrases are not apropos.
43. One should never generalize.
44. Eliminate quotations. As Ralph Waldo Emerson said, "I hate quotations. Tell me what you know."
45. Comparisons are as bad as cliches.
46. Don't be redundant; don't use more words than necessary; it's highly superfluous.
47. Be more or less specific.
48. Understatement is always best.
49. One-word sentences? Eliminate.
50. Analogies in writing are like feathers on a snake.
51. Go around the barn at high noon to avoid colloquialisms.
52. Who needs rhetorical questions?
53. Exaggeration is a billion times worse than understatement.
54. capitalize every sentence and remember always end it with a point

(Source: http://www.listsofnote.com/2012/01/fumblerules-of-grammar.html)

REFERENCES

Electric power substation (n.d.). *McGraw-Hill Dictionary of Scientific & Technical Terms, 6E. (2003)*. Retrieved from http://encyclopedia2.thefreedictionary.com/electric+power+substation.

Visible Earth (2015). *Aral sea*. EOS Project Science Office, NASA Goddard Space Flight Center. Retrieved from http://visibleearth.nasa.gov/view.php?id=78588.

Wallwork, A. (2013). *English for Research: Usage, Style, and Grammar*. New York: Springer.

Chapter 11

Do's and don'ts of technical writing

Vigorous writing is concise. A sentence should contain no unnecessary words, a paragraph no unnecessary sentences, for the same reason that a drawing should have no unnecessary lines and a machine no unnecessary parts. This requires not that the writer make all sentences short or avoid all detail and treat subjects only in outline, but that every word tell.

William Strunk Jr.
Elements of Style, 2012

Engineering demands objectivity, accuracy and precision and these qualities are also expected when writing in this field. The success of your report or thesis depends on your ability to convince your reader that your ideas are valid. Whether you are reporting a straightforward procedure or discussing your results, your writing should be easy to read and to understand.

This chapter is a concise guide to research writing. It provides guidelines on what you should do (the 'do's') and what you should avoid (the 'don'ts') to write your report so that it is clear and easy to read.

DO'S OF TECHNICAL WRITING

Because of its technical content, research writing is characterised by specialised vocabulary as well as by complex sentences that make it sound complicated. Novice writers often feel that they must adopt a similarly complicated style to sound intelligent and be taken seriously. This is neither true nor desirable. The most effective and readable style is one that is clear, concise, precise and consistent. The most important qualities are clarity and precision. Your goal in writing is to make it as easy as possible for the reader to understand exactly what you mean.

The following section provides guidance, first, on how to manage your writing (**view your writing as a process**) and, second, on how to write so that your report or thesis is easy to read and understand (**observe the 5 C's of technical writing**).

View your writing as a process

Approach your writing strategically by viewing it as a process comprising three stages: planning, drafting and revising. Set aside a realistic amount of time for each of the three stages.

The planning stage involves *gathering all necessary information* and *outlining*. Most of the information you need for your report should be in your research log (refer to *Chapter 8—Keeping research records* for further information on the research log). So review the entries in your research log and highlight those to include in your report.

Next, create a rough outline that details out the contents and organisation of each chapter of your report or thesis (refer to *Chapter 9—Starting to write* for specific instructions on how to create an outline). Write down the key points for each section of your report or thesis. Without an outline, it is easy to overlook important points or jump randomly from one idea to another. Remember that the outline is a tool to aid your organisation and it is fine to change it as you go along.

In *the drafting stage*, write from the outline and do not stop to revise or edit. You can do the editing after you have completed your draft. Drafting and editing at the same time will slow you down and interrupt your train of thought.

When you have a complete draft, you are ready to implement the final stage, revising and editing (refer to *Chapter 20—Revising and editing* for how to edit efficiently). Be prepared to work through two or three drafts, refining your work each time, before you are happy with the end result.

Observe the five C's of technical writing

In a report or thesis, write in a style that communicates the most information in the least reading time. The stylistic qualities required for this are the *five C's* of technical writing: *clear, concrete, concise, correct* and *consistent.*

Be clear

As Einstein said, 'Make everything as simple as possible, but not simpler'. Make your writing simple and clear by avoiding unnecessary complex words, phrases, and sentences and ambiguous language.

Prefer plain words over unnecessary complex words and phrases

Alley (1996) has pointed out that in technical writing many words are used which add no precision or clarity to the writing—only unnecessary complexity (pp. 83–90). Examples of unnecessarily complex words and their simpler (and clearer) substitutes are shown in Table 11.1.

Table 11.1 Examples of complex words and their simpler substitutes

Complex word	Simpler substitute
Axiomatic	Self-evident, unquestionable
Apprise	Inform, tell
Elucidate	Explain, clarify
Endeavour	Try, attempt
Impugn	Challenge, question
Modicum	A small amount
Obfuscate	Obscure, complicate
Peruse	Read, study
Propensity	Tendency, inclination
Purvey	Sell, supply
Desultory	Superficial, casual
Egregious	Extremely bad

Just as words in technical writing are often unnecessary complex, so too are phrases (groups of related words without a subject and a verb). One common source of unnecessary complexity comes from stringing a number of modifiers (adjectives and other nouns acting like adjectives) before a noun as in the sentence below:

'The decision was based on economical *fluid replenishment cost* performance.'

The long string of modifiers '*economical fluid replenishment cost*' in front of the noun '*performance*' confuses the reader. What exactly does the phrase 'economical fluid replenishment cost performance' mean? The revised sentence 'The decision was based on the cost of replacing the thermal oil' is simpler and clearer.

Avoid writing excessively long sentences

Longer sentences demand greater concentration from the reader. Excessively long sentences also introduce unnecessary complexity. Strive to keep your sentences between 15 and 25 words. If your sentence is longer than 25 words, have a second look at it and see if you can write it as two separate sentences.

Avoid overloading your sentences with too many different ideas

Your writing will be clearer and easier to read if you limit each sentence to expressing one main idea. If you have two different ideas, they are much more clearly conveyed when written as two separate sentences.

Avoid using ambiguous language

Ambiguity occurs when a word, a phrase or a sentence has more than one meaning and the reader cannot determine which is the intended meaning. Ambiguities arise from four specific sources: word choice, sentence structure, pronouns and punctuation.

- Ambiguities in word choice
  Many words in English have multiple meanings, for instance, consider the word 'affected' in the following sentence: 'The flow of water was affected.'
  'Affected' can mean that the flow of water was slowed down, sped up, or stopped.
  Replacing it with one of these more precise expressions would eliminate the ambiguity.

- Ambiguities in sentence structure
  When writers are not careful about the ordering of words and phrases in a sentence, ambiguities can occur. For example, the following sentence is ambiguous:
  'The memory module was lowered to the horizontal position that required testing.'
  The sentence is ambiguous because of the improper placement of the phrase *'that required testing'*. When this phrase is placed in a more logical position in the sentence as in 'The memory module that required testing was lowered to the horizontal position', the ambiguity is avoided.

- Ambiguities in pronouns
  An important principle of using pronouns is that the noun or verb which the pronoun refers to (the antecedent) should be clear. Unfortunately, this principle is commonly ignored, especially with regard to the use of the pronouns 'it' and 'this'. For instance, in the sentence below, there are many possible references for the pronoun 'it':
  'As the receiver presented the radiometer with a high flux environment, it was mounted in a silver-plated stainless steel container.'
  Given the number of possible references *(receiver, radiometer, environment)* for 'it', the sentence is much clearer if the writer repeats the noun 'radiometer' (the intended reference): 'As the receiver prevented the radiometer with a high flux environment, the radiometer was mounted...'
  To avoid confusion, it is also advisable to name whatever 'this' refers to immediately after 'this' (i.e. 'this phenomenon,' 'this principle,' 'this variation'). Note how much clearer the following sentence is because 'this assumption' is used rather than if the pronoun 'this' is used alone:
  'Tectonic burial by thrusting is believed to occur rapidly. This assumption, however, is difficult to test.'

- Ambiguities in punctuation
Absence or improper use of punctuation is a common source of ambiguity. In the sentence below, punctuation is needed to keep the sentence from being misread (Alley, 1997).

'Neat methanol, neat ethanol, methanol and 10% water and ethanol and 10% water were examined in this study.'

It is unclear from the above sentence how many fuels were examined. To eliminate the ambiguity from the sentence, the writer should use a colon and a comma before the final item in the series:

'In this study, the following four fuels were examined: neat methanol, neat ethanol, methanol with 10% water, and ethanol with 10% water.'

Note: For other common errors in punctuation and how to avoid them, refer to *Chapter 10—Grammar, punctuation and word usage guide*.

Be concrete (precise)

Choose the right words

Communicate exactly what you mean by choosing the right words. Do not choose the word 'weight' when you meant 'mass' as these are technical terms with specific meanings. Many ordinary words have specific meanings as well. For instance, 'comprise' and 'compose'. The sentence 'Water comprises hydrogen and oxygen' is imprecise because 'comprise' means 'to include'. A precise way to write this sentence is 'Water is composed of hydrogen and oxygen'.

For other word pairs which are commonly confused, see the section 'Pairs of words commonly confused' in *Chapter 10—Grammar, punctuation and word usage guide*.

Avoid imprecise or inexact statements

When you write, make sure each statement conveys the exact meaning you intend and not another. Examples of statements with imprecise meaning are given in Table 11.2.

Table 11.2 Examples of imprecise statements

Imprecise statement	Comment
The readings were reasonably accurate.	Use numbers if possible, for example: *Sixty percent of the readings were accurate.*
The gear failed because of an excessive number of load cycles.	Provide an exact figure in place of *excessive number*. If this is not possible, provide an estimate.
There were millions of data points to be examined.	Avoid exaggerated statements or inflated claims as they reduce the writer's credibility.

Be mindful of word connotation

Words have dictionary meaning (denotation) as well as connotation. A word's connotation is its associated meanings which can be negative or positive. Avoid using a word with a negative connotation when a neutral or positive connotation is intended. Similarly, avoid using a word with a positive connotation when a negative or neutral connotation is desired as in the sentence below:

'The turbulence in the flow enhanced the drag by more than 30%'.

In the example, the word 'drag' is an undesirable quality. Thus the writer should have chosen a verb with a neutral or negative connotation instead of 'enhance' which has a positive connotation. A better word choice would have been either 'increase' (neutral connotation) or 'exacerbate' (negative connotation).

Be concise

When writing, make every word count. Get rid of words and phrases which serve no purpose other than to make your writing long-winded. You can make your writing concise in the following ways:

Eliminate redundancies

Redundancies in writing are needless repetition of words within a sentence. Redundancies either repeat the meaning of an earlier expression or else state what is obvious. For instance:

'The aluminium metal cathode became pitted during the glow discharge.'

The word 'metal' is redundant after the word 'aluminium'. Table 11.3 lists some common redundancies in technical writing. The words in parentheses can be omitted.

Eliminate meaningless phrases

Writers often use phrases which offer no information to the readers. For instance, the phrase 'it is interesting to know' in the sentence 'It is interesting to note that over 90 incidents of satellite fragmentations have produced

Table 11.3 Examples of redundancies

(Already) existing	Cylindrical (in shape)
(Completely) eliminate	Mix (together)
(Uniformly) homogenous	Repeat (again)
(Basic) fundamentals	Period (of time)
(Continue to) remain	Never (before)
(Past) experience	Start (out)

over 36,000 kilograms of space debris' is superfluous. Table 11.4 shows some common examples of meaningless phrases which should be omitted in your writing.

Replace wordy phrases with shorter alternatives

Whenever possible, replace wordy phrases with more concise equivalents as in the examples given in Table 11.5.

Be correct

Use formal language

Reports, theses and dissertations are formal documents. This means you should avoid:

- Slang such as *dicey, awesome, and iffy, hassle.*
- Colloquial language such as using *like* when the proper word is such as. Write '*such as* platform for chip installation and not *like* platform for chip installation'.

Table 11.4 Examples of meaningless phrases

As a matter of fact	It should be pointed out that
It is interesting/significant/noteworthy that	The fact that
In the course of	We might add that
In the presence of	Notwithstanding the fact that
It is unnecessary to state that	It goes without saying that

Table 11.5 Examples of wordy phrases (to be avoided)

Wordy phrase	Concise equivalent
At this point in time	Now; currently
During the course of	During
Has the ability/potential to	Can
There is no doubt that	Undoubtedly
In the event that	If
Exhibits the ability to	Can
For the purpose of (investigating)	To (investigate)
In order to	To
In spite of the fact that	Though/although
In view of the fact that	Because
It would appear that	Apparently
Owing to the fact that	Since
The reason why is that	Because

Contractions, such as *don't, can't, didn't, wasn't, it's* and so on.
Personal pronouns such as *I, me* and *you.*
Emotive language such as luckily, unfortunately, surprisingly, thankfully.

Use correct grammar

Being grammatically correct means first of all using the correct tense. It is best to use the past tense when writing about what you did. But be aware of confusing the use of the simple past and the simple present because this will affect the meaning of your sentence. Among its uses, the present simple is used to express habitual actions. Guidance on using tenses for your report or thesis is given in *Chapter 10—Grammar, punctuation, and word usage guide.*

Write well-formed sentences

Well-formed sentences are sentences which conform to grammatical rules of English. The following are the three criteria for well-formed and grammatical sentences:

1. The sentence has *a subject.*
   Electrospinning [subject] can be used with different synthetic as well as natural origin polymers.
2. The sentence has *a complete verb.*
   MSCs *were cultured* [complete verb] in a petri dish.
3. The sentence conveys *a complete thought.*

Compare the following two sentences:

a. Although the 'corridor' for ships travelling to and past Singapore is well defined.
b. Although the 'corridor' for ships travelling to and past Singapore is well defined, this 'corridor' is not closely monitored.

Sentence (a) is unclear and ungrammatical as it is actually a dependent clause disconnected from the main clause (*this 'corridor' is not closely monitored*). Sentence (b) conveys a complete thought as the dependent clause (*Although the 'corridor' for ships travelling to and past Singapore is well defined*) is linked to the main clause (*this 'corridor' is not closely monitored*).

Note: Refer to *Chapter 10—Grammar, punctuation, and usage guide* for common grammatical errors and tips on how to avoid them.

Be consistent

Consistency is an important requirement in technical writing especially in using key words, in spelling, and in the format for units.

Table 11.6 Examples of differences between British and American spelling

British	American
Analogue	Analog
Colour	Color
Defence	Defense
Organise	Organize
Analyse	Analyze
Fulfil	Fulfill
Centre	Center
Manoeuvre	Maneuver

It is advisable to use the same word (e.g. network) to refer to a key concept rather than use a number of synonyms ('system', 'configuration' etc.) merely to add variety as this may confuse the reader.

In spelling, use either British or American English spelling but not both in the same document. Examples of difference in spelling between those two are shown in Table 11.6 (note differences in spelling different word endings and vowels).

To maintain consistency in spelling, select the variety of English (British English or American English) for your computer before you start your writing project. The computer will then auto-correct should you use the American spelling of a word if you have selected British English and vice versa.

You should also use a consistent format for units. For example do not randomly use 'V', 'Volts' and 'volts' in your document.

DON'TS OF TECHNICAL WRITING

The following are habits to avoid to ensure that your writing is clear and correct.

Don't use jargon unnecessarily. Use specialised terms only if they are understood by your readers. Otherwise, these terms will be 'jargon' to your readers.

Don't combine words to make up new words. Don't use words that are not in the dictionary (e.g. *pathloss, basestation* etc.).

Don't start a sentence with a number. Write the numeral as a word (e.g. *Ten profiles were evaluated*, and not *10 profiles were evaluated*).

Don't start a sentence with an abbreviation. Write out in full if it is the first word of the sentence (e.g. *Figure 10 shows ...*, and not *Fig. 10 shows...*).

Don't use absolutes. Avoid extreme words such as 'always' and 'never'. Whenever you use these words, you invite your reader to look for exceptions. And if your reader does find them, the credibility of your work and of you will suffer.

Don't assume a universal format. In engineering and science, each discipline (or sub-discipline) has its own format conventions.

INTERESTING FACTS

Gunning Fog Index

The Gunning Fog Index is one of the best known tools for determining how difficult it is to read and understand written texts in English. It is a weighted average of the number of words per sentence, and the number of long words per word. The index estimates the number of years of formal education needed to understand a text on first reading.

Find out how readable your writing is using the Gunning Fog Index at http://gunning-fog-index.com/.

Readability Check in Microsoft Office Word.

Microsoft Office Word provides readability scores of your Word document according to Flesch reading ease and Flesch-Kincaid Grade Level tests.

Find out how to use this feature in Microsoft Office Word by clicking the Help button and searching for 'readability'.

REFERENCE

Alley, M. (1996). *The Craft of Scientific Writing* (3rd ed.). New York: Springer-Verlag.

Chapter 12

Strategies for writing a good report or thesis

The best dissertation is a finished dissertation.

Kate Drowne
Writing Successful Theses and Dissertations, 2007

What is a good report or thesis? A good report or thesis has the following characteristics:

- It explains the full motivation for the work.
- It explains the process through which the knowledge and results were acquired.
- It provides sufficient details to make it interesting to the reader.
- It sets out the limitations of the work and provides recommendations for future research.
- It contributes to the literature.
- It is grammatically correct and has a logical layout.

Writing a good report or thesis requires careful planning at the start and persevering until the report or thesis is completed. Strategies for helping you to accomplish this task are described in the following.

LEARN FROM OTHERS

Reading the reports and theses written by other students in the same institution will give you a good idea of your institution's requirements for a report or thesis. Reading reports and theses in the same field or on a similar topic from other institutions will help you understand how others have approached writing a report or thesis in your field or topic. You should also not restrict yourself to reading just one report or thesis but read several so that you can get a better sense of what makes a good report or thesis. A good report or thesis should have clear objectives and research methodology, the results and findings should be correct and accurate, and the writing should be clear, concise and objective.

START EARLY

Writing a good report or thesis is similar to cooking a good dish: it requires high-quality ingredients. For a report or thesis, the ingredients are the materials. Without high-quality materials, no amount of good writing will produce a good report or thesis. These materials include literature review, methodology, experiment details, results, data, charts, figures, tables and references. The analogy between cooking a good dish and writing a report or thesis then ends as writing a report or thesis, unlike cooking, is not a linear process but a recursive process. The quality of a report or thesis will improve with each revision and you need to set aside sufficient time for revision. However, a research schedule is usually tight and you may have to start writing the thesis before all the materials are ready.

Writing the thesis should start right at the beginning of the project (see *Chapter 5—Writing a research project proposal, Chapter 8—Keeping research records and Chapter 9—Starting to write*).

WRITE THE FIRST DRAFT QUICKLY

It can be a daunting task to write the first draft of a report or thesis even if you have the materials described in Chapters 5, 8 and 9 to help you. Most students find it hard to write the first chapter, the *Introduction*, and spend a disproportionately large amount of time writing it. This experience may explain the fear that many students have of writing a report or thesis and their tendency to put it off until the last possible moment.

The task can be made less daunting and more efficient by adopting a simple writing strategy for the first draft. This strategy involves dividing the report or thesis into three main parts: (1) your work, (2) others' work and (3) summation. The chapters in the report or thesis that are *your work* are *Methods and Materials* (or *Methodology*) and *Results and Discussion*. The chapter in the thesis that is *others' work* is the *Literature Review*. The chapters in the thesis that are the *summation* are the *Abstract* (or *Summary*), the *Introduction* and the *Conclusion*.

An effective strategy in writing is to start with the familiar and then move on to the less familiar. The part of the report or thesis that is most familiar to you is your own research on which you have spent the most time. Therefore, if you start your first draft by writing about your work, the *Methods and Materials* and *Results and Discussions*, you are off to a flying start (see *Chapter 14—Writing the methods and materials and chapter 15—Writing the results and discussion*). You should be able to write up your work in a relatively short time if you adopt the strategy of starting with the familiar. After the first draft of the *Methods and Materials* and *Results and Discussion* chapters have been written, they can be revised to make them consistent with each other.

Once your own work has been written up, you can then move on to write the *Literature Review* chapter *(others' work)*. Writing the Literature Review after you have written the *Methods and Materials* and *Results and Discussion* chapters will help to ensure that you include only literature that is directly relevant to your work. You will also be able to appraise the literature critically from the perspective of your own work. While writing the *Literature Review* chapter, you may want to revise the earlier chapters of your work, especially the *Results and Discussion* chapter, as you may want to cite some of the publications reviewed.

Only after you have drafted the chapters of *your work* and *others' work* should you start writing the *summation* chapters, which are the *Introduction, Conclusion* and *Abstract,* in that order. The *Introduction* comprises the *background, objectives and scope* (see *Chapter 13—Writing the introduction and literature review*). The *Conclusion* sums up the findings and states if the objectives are fulfilled (see *Chapter 16—Writing the conclusion*). Therefore, the writing of the *Conclusion* should immediately follow the *Introduction.* The *Abstract* is basically a summary of the report or thesis (see *Chapter 17—Writing the abstract and front matter*). Once the *summation* chapters are written, you can then revise them to maintain consistency and coherence. The strategy in writing the first draft of a report or thesis is summarised in Figure 12.1.

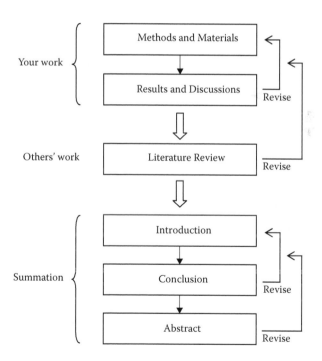

Figure 12.1 Strategy for writing the first draft.

Once the first draft of the report or thesis is written, you will have essentially a complete report or thesis. Revisions can be done sequentially, from the first chapter to the last chapter of your report or thesis, to ensure that the chapters are all logically connected. The writing of a good report or thesis lies in the number of revisions that are made to 'polish' it. You should aim to revise your report or thesis at least once, preferably more. For more help with revising and editing your report or thesis, read *Chapter 20—Revising and editing.*

USE A TEMPLATE

Use the in-built productivity tools in a word processor such as Microsoft Office Word to help you write more efficiently. One such productivity tool is the template file. The template file is a collection of styles and format settings that helps you to be consistent when you write using word processing software. You may want to check with your supervisor or institution if a report or thesis template file is available from your institution. If not, you can either choose to use an existing template file or you can create your own.

However, it is easier if you use one of the existing template files and modify it to suit your institution's requirements (see *Chapter 22—How to create a good layout*—for information on creating a template file). Many report and thesis template files can be found on the Internet. In the template file, the page size, margins, header, footer and pagination for a page are pre-defined and there is a particular style for the title, headings, first paragraph, body text, sub-headings, table' captions and figure' captions. In Microsoft Office Word, headings and sub-headings are labelled as Heading 1, Heading 2 and so forth. By applying the style to the headings, sub-headings, table' captions and figure' captions, you can generate the *Table of contents*, the *List of figures* and the *List of tables* quickly. These parts of the thesis are tedious to generate manually but they are essential to your thesis.

Assuming that you have applied *Heading 1* to your chapter titles, *Heading 2* to your section headers, *Heading 3* to your sub-section headers and *Heading 4* to your sub-sub-section headers and so forth, you can generate the *Table of Contents* in Microsoft Office Word as follows:

1. Go to the page where you want to place the Table of Contents.
2. Select the *References* tab in the top menu.
3. Click *Table of Contents* under the *Table of contents* group.
4. Select the *Table Style* that you want.

Similarly, you can generate the List of Figures and List of Tables in Microsoft Office Word automatically. Before you can generate the List of Figures or List of Tables, you need to do the following steps:

1. Go to the location of the figure caption (usually at the bottom of the figure) or table caption (usually at the top of the table) and, select the *References* tab in the top menu.
2. Click *Insert Caption* in the *Captions* group and a Caption dialog box appears. In the *Options group*, select *Figure* in the drop-down *Label box* for figure caption (see Figure 12.2) or select Table (in italics) in the drop-down Label box (in italics) (see Figure 12.3) for table caption.
3. Type the caption for the figure after Figure 1 or for the table after Table 1 in the *Caption* box and click *OK*.
   To generate the List of Figures: 1. Go to the page where you want to place the List Figures.
4. Select the *References* tab in the top menu and select *Insert Table of Figures*. The List of Figures will be automatically generated.

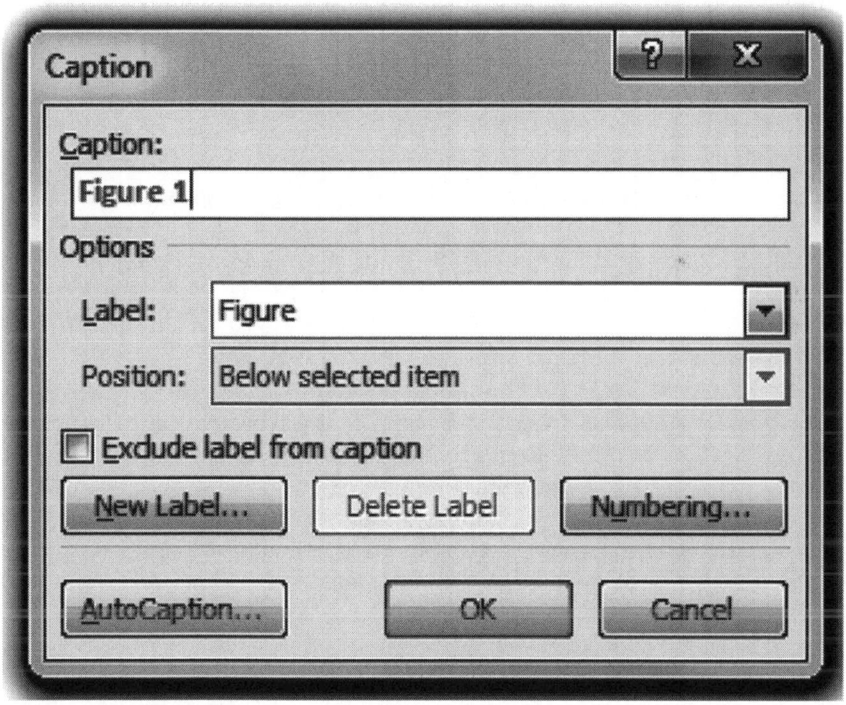

Figure 12.2 Caption dialog box for figures.

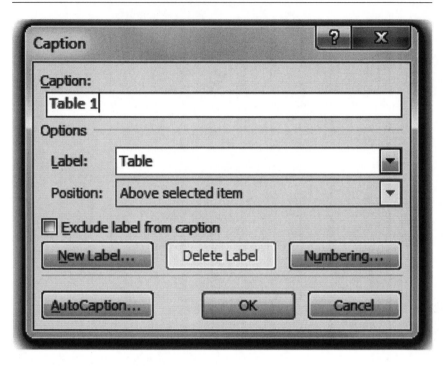

Figure 12.3 Caption dialog box for tables.

To generate a List of Tables:
1. Go to the page where you want to place the List of Tables.
2. Select the *References* tab in the top menu and select *Insert Table of Figures*. The List of Tables will be automatically generated.

REVISING AND CHECKING FOR PLAGIARISM

A well-written report or thesis is the result of several rounds of revising and editing (see *Chapter 20—Revising and editing* for more detailed advice). Consider the following when revising your draft:

1. Are there any parts that can be omitted without loss of essential information? These could be text, figures or charts.
2. Are there any sentences and paragraphs that can be re-written more clearly and concisely?
3. Are there any spelling, grammatical and punctuation errors?
4. Is there any information that you have omitted in a citation?

Next, you will want to check if you have accidentally reproduced text from other publications. If your institution subscribes to software such as

Turnitin, Grammarly, or PlagTracker, you can run your thesis through the software to check for accidental plagiarism. For further information on plagiarism, read *Chapter 19—Using sources and avoiding plagiarism.*

The strategies mentioned above are basic strategies for writing a good report or thesis. If you come across other strategies that you think are helpful, add them to the strategies above to ensure that your report or thesis is written well and on time.

INTERESTING FACTS

Numerous paper awards are given out regularly at conferences, by professional bodies and by journal publishers. Find out the paper awards in your field of research and read the award-winning papers to understand how such papers are written. Examples of such awards are as follows:

The Best Paper Award of the IEEE Signal Processing Society is given to a paper appearing in one of its solely owned transactions or the *Journal of Selected Topics in Signal Processing.*

The Blackall Machine Tool and Gage Award is given to the best paper on the design or application of machine tools, gages or dimensional measuring instruments by the American Society for Mechanical Engineers (ASME).

The Thomas A. Middlebrooks Award is given to the author(s) of a paper published by the American Society for Civil Engineers for its merit as a contribution to geotechnical engineering.

Chapter 13

Writing the introduction and literature review

The measure of greatness in a scientific idea is the extent to which it stimulates thought and opens up new lines of research.

Paul Adrien Maurice Dirac
The Scientific Work of Georges Lemaître, 1968

The *Introduction* is the first chapter of the report or thesis and its chief function is to familiarise the reader with your research problem and the relevant literature. The literature review may be a section within the *Introduction* or written as a separate chapter following the *Introduction*. The *Literature Review* as a separate chapter is a more in-depth counterpart of the literature review section within the *Introduction*. In a thesis, the *Literature Review* as a separate chapter is usually the norm.

This chapter provides guidance on writing the *Introduction*. First, the components of an introduction will be described. This will be followed by a discussion of the common logical missteps committed by writers and, finally, the language features that you can use to enhance your writing will be highlighted.

COMPONENTS OF AN INTRODUCTION

The *Introduction* essentially comprises the following three basic components:

1. Establish key ideas and findings of research literature and/or real-world context relevant to your research topic.
2. Present a research problem.
3. Announce how you will specifically address or fix the research problem.

These components are further broken down into steps as shown in Writing Guidelines 13.1. The components and steps are based on a synthesis of the research findings on how introductions are written from the work by Swales (1990, 2004), Bunton (2002) and Samraj (2004, 2008) as well as our analysis of engineering reports and theses.

Writing Guidelines 13.1 Components of *Introduction/Literature Review*

Component 1: Establish key ideas and findings of research literature and/or real-world context relevant to your research topic

A. Indicate importance/recency of relevant literature (very frequently used)
B. Indicate claims based on research (citing and reviewing research) or common knowledge (very frequently used)
C. Define terms and constructs in your research that are not frequently used (infrequently used)

Component 2: Portray a research problem of uncertainty

One or more of the following statements can be used:
A. Argue for a real-world problem or need (very frequently used)
B. Argue that no research was conducted on your topic (frequently used)
C. Argue that scant research was conducted on your topic (very frequently used)

Component 3: Announce how you will specifically address or fix the research problem

Two or more of the following statements are used:
A. State the aim/purpose/objective of your research (very frequently used) and/or describe theoretical and/or real-world contributions of your research (frequently used)
B. State research question or hypothesis of your research (frequently used)
C. Summarise the research design/method of your research (frequently used)
D. State key findings of your research (infrequently used)
E. Indicate scope/parameters of your study (infrequently used)
F. State explicitly the significance or value of your research (infrequently used)
G. Outline the structure of all sections subsequent to the introduction (very frequently used)

In Component 1, share with the reader the key concepts and findings from previous research that are related to your research topic. In other words, you need to situate your research in the wider context of research and/or the real world. In a way, Component 1 reflects the expectation of the reader that your research topic is developed from links to previous research and not formulated in vacuum or isolation. There are three steps in Component 1:

Step A: Statements about the importance or centrality of your research topic in research and/or the real world. You can claim importance directly or indirectly. You can claim importance directly by stating the importance or significance of your research topic (e.g. 'Indeed, the search of natural compounds and management methods alternatives (or complements) to classical pesticides and fungicides has become an *intense* and *productive* research field'). Note the keywords in italics which explicitly declare the importance of the research area. Alternatively, you can claim importance indirectly by stating the high frequency and recency of the research done (e.g. 'In *recent* years there has been a *significant, resurgent interest* in renewable energy sources').

Step B: Stating a knowledge claim which can be an established fact or a controversial finding. You may or may not substantiate a claim with an in-text citation (see *Chapter 18—Referencing, in-text citations*). If there is no in-text citation to back up a claim stated in your report or thesis, it must be supported by common knowledge and is, therefore, unlikely to be challenged by the reader. For example, a claim that states that all fossil fuels are finite resources is a fact which does not need to be supported by cited studies because it is deemed to be common knowledge by academic readers. If the claim is controversial, support it by citing empirical sources. For example, a claim that alleges that nuclear energy is the best sustainable energy source is controversial and needs to be supported by cited scholarly sources.

Step C: Definitions of technical terms and theoretical constructs in your research.

Component 2 is another indispensable component in the *Introduction and Literature Review*. It usually follows Component 1, after the establishment of the most relevant research/real-world context to your research topic. However, Components 1 and 2 can be intertwined with statements belonging to either part.

Component 2 sets the stage for the justification of your research topic as a useful study in the wider context of research and/or the real world by portraying a research problem of uncertainty—this specific research uncertainty is your research topic. Your research topic should be narrowly framed or defined, particularly in a report or thesis with a strict word limit. Notably, the language used in portraying a research problem is usually negative and contains words and expressions such as *suffer from weaknesses, can work only, none of, no, not, fail and inconclusive.* The claim that your research topic is justified as previous research fails to establish certainty can be done in three steps:

Step A: In this step, you 'argue for a real-world problem or need'. This is the most common approach used by engineering writers—expectedly so as engineering is closely applied in the real world. An example sentence is, 'To reduce heavy reliance on herbicides and fungicides, there is *a need* to move to low-input sustainable agriculture as a component of integrated weed and fungi management.' Step A is usually accompanied by knowledge claims (Component 1, Step B) which furnish key supporting details underlying the real-world problem or need.

Step B: Here you argue for a research gap that is neglected which your study will fill. An example sentence is, '*However, no* information is available about its antifungal activity.' Negative language is marked in the use of *no* and implied in the logical connector, *however*.

Step C: You argue for a problem of research insufficiency or inadequacy. Before a finding can be established as a fact, empirical science requires that a critical mass of experimental studies corroborate it. Insufficient or inadequate research can be argued on the basis that this scientific consensus has not been reached or achieved. An example sentence is, 'While there have been numerous studies examining the biomechanics of typing for body postures, head postures, arm postures, and wrist postures, there have been *only three* studies (Baker et al., 2007; Sommerich et al., 1998, 1996) that have looked at the postures and actions of hands and digits during standard keyboard use.'

Component 3 is the concluding component that explicitly states how your research addresses or fixes the research problem stated earlier. This component is accomplished in seven steps, steps A–G explained below.

Step A: Statements of your research aims, objectives or purposes, or a description of theoretical and/or real-world contributions of your study. An example of stating your research aims, objectives or purposes is, 'This study compared the kinematics of the digits during typing on a standard keyboard configuration and an ergonomic keyboard configuration *to determine* if an ergonomic keyboard configuration reduced the digit postures and motions that are hypothesised to be risk factors for MSD-UE.' The keywords (to + verb infinitive) in italics indicates the purpose of the study. An example of describing theoretical and/or real-world contributions of your study is, 'The present work *extends* the use of the last model to asymmetric, body-vortex flow cases, thus *increasing* the range of flow problems that can be investigated.' Note that the language used is positive, as indicated by the words in italics.

Step B: Statements of research questions or hypotheses are made. Research hypotheses are hedged statements* of best guesses of your research problems based on what past findings can glean. An example of research hypotheses is (note the cautious words in italics), 'The following two hypotheses were examined: (1) participants using the ergonomic configuration *would* demonstrate significantly more neutral joint angles, slower joint angle velocities and reduced joint angle accelerations at the metacarpophalangeal (MCP) joint (flexion/extension and abduction/adduction) and proximal interphalangeal (PIP) joint (flexion/extension) compared to using the standard configuration; (2) participants using the ergonomic configuration *would* demonstrate greater hand displacements compared to using the standard keyboard.'

Step C: Summary of the research design especially when the research method is the central novelty of a research project.

Step D: Component 3 can also include a statement on the key findings of your research, such as 'The estrone-adsorbing capability of

---

* 'Hedged statements' are statements of uncertainty.

nylon membranes, as reported in this study, *can augment* their solids removal functions by providing simultaneous treatment for source water that contains elevated levels of estrone. A *unique* feature of this method is that it can operate at low pressures as it is driven by a chemisorption mechanism and *does not* rely on size exclusion.' Note the words in italics which signal the importance of the findings.

Step E: The parameters or scope of your research study can be indicated as a part of Component 3.

Step F: This step is a bold proclamation of your study's research or real-world worth such as 'The synthesis strategies of both mesoporous carbon and Pt/GMPC *are facile and effective.*'

Step G: The last step of Component 3 is a content primer which briefly summarises the contents of subsequent sections of the report, such as 'A brief summary of leading approaches to the first two questions *will* be described, *followed by* a general description of how proposed approaches employing nanoscale structures are capable of answering the third question.' The modal verb 'will' and the verb + preposition phrase 'followed by' indicate the sequence of information after the introduction.

STRATEGIC USE OF STEPS IN THE DIFFERENT PARTS OF AN INTRODUCTION AND LITERATURE REVIEW

In writing the first draft of your research proposal, literature review or introduction of your report or thesis, the three components described above should be presented briefly. Scope-wise, Component 1 is most general in covering relevant literature, followed by Component 2 which narrows in on a problem of uncertainty which is your research topic. Component 3 specifically spells out how you will address your research topic. In Component 1, Steps A and B are frequently used and should be present in your writing. Step C (supplying technical definitions) is less frequently employed.

In Component 2, the number of steps used depends on the word length of your report or thesis. If your report or thesis is short, say less than 3,000 words, it is likely that you will choose only one of the three steps as your research problem. If you have more word space, you may use more than one step in Component 2.

Component 3 comprises an assortment of steps from which you can pick and choose as a writer. Note that the most important and frequently used steps in Component 3 are Steps A (state the aim/purpose/objective of your research) and G (outline the structure of all sections subsequent to the introduction). Writing Guidelines 13.2 presents an analysis of the contents of two sample introductions. The component as well as the step within the component that each sentence shows is indicated in brackets.

Writing Guidelines 13.2 Content analysis of two sample introductions

Example 1: Abstracted from Cheng and Leong (2014)

Determination of small strain parameters like shear modulus (G) and damping ratio (ξ) are frequently carried out with advanced geotechnical tests like the resonant column, the cyclic simple shear and the cyclic triaxial tests **(1B)**[a]. Both the cyclic simple shear and cyclic triaxial tests have been included in testing standards and hence been widely used in the industry **(1B)**. Another popular test to determine small strain parameters is the wave propagation test using bender/extender elements **(1B)**. The bender/extender elements are able to determine small strain parameters at strains smaller than the resonant column and cyclic triaxial tests (Das, 1993) **(1A)**. The wave propagation test has the further advantage of being incorporated into testing apparatuses like triaxial apparatus, resonant column and simple shear apparatus **(2A)**.

While most research had focused on the determination of the stiffness modulus especially the shear modulus, material damping ratio is seldom investigated **(2B)**. This paper explores two methods of determining damping ratio using the bender element tests **(3A)**. The two methods are the Logarithmic Decrement and the Spectral Ratio Method **(3E)**.

Example 2: Abstracted from Wijaya and Leong (2014)

Climate and groundwater table has a major role in causing a soil to be saturated or unsaturated (Fredlund and Rahardjo 1993) and therefore cause the soil to undergo swelling or shrinking (Kim et al. 1992) **(1A)**. Jones and Holtz (1973) reported that the cost of damages to houses, buildings, roads and pipelines due to shrinking and swelling of soils caused is twice as much as the cost of damages from floods, hurricanes, tornadoes and earthquakes in US **(1A)**. Therefore considerable research efforts have been placed on predicting shrinkage and swelling of soils **(1A)**. Several factors which have been considered to influence the shrinkage properties of the soils are the soil structure, the initial water content, the type of clay mineral, the clay content, the organic matter content, the kind and concentration of the cation of the pore water and the drying conditions (Japanese Geotechnical Society 2009; Umezaki and Kawamura 2013) **(1B)**.

Nelson and Miller (1992) listed at least four techniques that are used to predict the shrinkage and swelling behaviors of soils which are the consolidometer test without suction measurement (ASTM D3877, 2008; ASTM D4546, 2008; ASTM D4829, 2011), consolidometer test with suction measurement (Fredlund and Morgenstern 1976), constructing the shrinkage curve (Hamberg 1985) and empirical procedures (Schneider and Poor 1974; Van der Merwe 1964) **(1B)**. Empirical procedures may not be accurate as they depend on the assumptions which were used to develop the procedures while consolidometer test without suction measurement gives only the maximum swelling/shrinkage potential **(2E)**. When suction measurement is performed, it is possible to obtain the change in swelling/shrinkage due to the change in matric suction **(2C)**. However, it requires additional equipment such as a high-capacity tensiometer (Guan and Fredlund 1997; He et al. 2006; Ridley and Burland 1993; Ridley and Burland 1999; Tarantino and Mongiovi 2002) **(2A)**. Constructing the shrinkage curve is the easiest method to predict the amount of swelling and shrinkage and it requires the least equipment **(2A)**. Therefore, the objective of this paper is to construct the shrinkage curves of different types of soft soils **(3A)**. Several shrinkage curve equations which are considered to have high accuracy were used to represent the shrinkage curves data **(3C)**. The comparison between different shrinkage curve equations is then used to give recommendation on the most suitable equation to be used **(3C)**.

[a] Bracketed term at the end of each sentence indicates a step, for example **(1B)** indicates step B of Component 1, **(2D)** indicates step D of Component 2 and so on as shown in Writing Guidelines 13.1.

COMMON LOGICAL PITFALLS IN AN INTRODUCTION

It is important that you substantiate every claim that you made which is not based on common knowledge with in-text citations. In other words, a claim that is potentially contestable should be backed up by empirical evidence.

If you support a general claim with one citation, this may not be deemed sufficient evidence to reduce serious objections from the reader. For example, if you made a debatable claim that girls are better than boys in mathematics and backed it up with one study, the evidence may be deemed insufficient because of the controversial nature of the claim. Readers may infer that this is a case of sweeping generalisation or overgeneralisation based on only one study's findings. Also, if the reader is aware of counter-evidence from other studies, you may be perceived as cherry-picking evidence to support a biased view. Such pitfalls can be avoided if you conduct a thorough and critical literature review.

SOME KEY LANGUAGE FEATURES OF AN INTRODUCTION AND LITERATURE REVIEW

Verb tenses

Verb tense is traditionally used in the temporal sense to locate an event in time, past or present. However, verb tense in the *Introduction* can also be used rhetorically, or persuasively. In Component 1, present tense and present perfect tense are used to indicate that a claim (Steps A and/or B) is a 'generally accepted truth'. When you cite and review research in Component 1 Step B, the present perfect tense and the past simple tense are used. If you are describing research on a specific phenomenon that is collectively conducted over time until the present, then the present tense should be used. If you are referring to a study's research objectives and/or findings, the simple past tense should be used (Bitchener, 2010). See *Chapter 10—Grammar, punctuation and word usage guide* for a more detailed explanation of tenses.

Evaluative words

Evaluative words are used to indicate your position or evaluation of an idea. Cautious words such as *may, can, and probably* indicate a level of uncertainty towards what you have read. Hyland (2000) states that cautious words are used for the following reasons:

- You are uncertain about what you said or read.
- You want to make clear that you are only presenting your opinion.

- You are certain about what you are claiming but want to be seen as modest or show deference to your reader.
- You acknowledge that readers may have reservations about your claim.

Although a cautious word weakens a claim, the weakened claim is less vulnerable to objections because the reader cannot deny your weakened claim. However, even after you have weakened a claim, you may need to elaborate on one or two alternatives or objections and evaluate them, especially if you anticipate that your reader will require more information.

WRITING THE LITERATURE REVIEW AS A SEPARATE SECTION OR CHAPTER

In a thesis or report, the *Literature Review* is usually a section or a chapter on its own. Sometimes, a journal paper may also have a literature review as a separate section. The *Literature Review* on its own is more informative than writing the literature review as part of the *Introduction*. The *Literature Review* serves to provide the link between your research and previous research. Most importantly, the literature review enables you to gain a perspective of your research topic, and helps you to avoid duplicating research efforts and to identify unforeseen problems. Therefore, the contents of a *Literature Review* should include the following:

1. Relevant account of all research and theories relevant to your topic
2. Historical development of your topic if your topic is not entirely new
3. Links between the various research areas and identification of the research gaps
4. Summary of all methods, analyses and techniques relevant to your topic
5. Identification of how your research can contribute to the existing knowledge

A good literature review needs to be critical, that is, you have to take a stand on the issues. It should not just be a description of past research work but an analysis of past research and how it is linked to your research topic. Any criticism needs to be substantiated by a balanced evaluation, it should not be cherry-picking of research work that is 'favourable' to your view. An unbalanced or biased literature review can be easily sensed by the reader.

The literature review is an ongoing process during your research up to the point in time that you are writing it up. There has been anecdotal accounts of PhD candidates who have to redo their research because someone else has beaten them to it. It is thus important to keep abreast of the latest developments in your research topic. Therefore, make literature review, both reviewing others' work and writing the literature review, a regular part of your research work. The following are tips on writing your literature review:

- Identify keywords in your research topic.
- Use the keywords to identify publications most relevant to your research topic. You can start with about 10 publications.
- Read the abstract of the identified publications and zoom in to the information that is most relevant to your topic.
- Take notes and compile the information as a summary list in chronological order.
- From the summary list, identify common ideas and form linkages between the publications.
- Write the first draft of the literature review based on the 10 publications.
- Add on to the literature review over time as you read more publications as your research progresses.

CHECKLIST FOR INTRODUCTION AND LITERATURE REVIEW

Do your introduction and literature review include the following:

- ☐ Key findings of studies situated within the research field of your topic?
- ☐ Arguments for a research problem of uncertainty which you will address in your study?
- ☐ Statement of your research objective or aim?
- ☐ A list of research questions which your study aims to address?

INTERESTING FACTS

Louis de Broglie's 1924 doctoral thesis titled 'On the Theory of Quanta' consists of only 70 pages, but it revolutionised quantum theory by theorising that electrons and all other matter have wave-like properties. After Clinton Davisson and Lester Germer confirmed de Broglie hypothesis empirically in 1927, de Broglie received the Nobel Prize for Physics in 1929.

REFERENCES

Bunton, D. (2002). Generic moves in PhD theses introductions. In J. Flowerdew (Ed.), *Academic Discourse* (pp. 57–75). Harlow: Longman.

Hyland, K. (2000). *Disciplinary Discourses: Social Interactions in Academic Writing.* London: Longman.

Samraj, B. (2004). An exploration of a genre set: Research article abstracts and introductions in two disciplines. *English for Specific Purposes* 24: 141–156.

Samraj, B. (2008). A discourse analysis of master's theses across disciplines with a focus on introductions. *Journal of English for Academic Purposes* 7: 55–67.

Swales, J. (1990). *Genre Analysis: English in Academic and Research Settings.* Cambridge: Cambridge University Press.

Swales, J. (2004). *Research Genres: Explorations and Applications.* Cambridge: Cambridge University Press.

ADDITIONAL READING

Note that the quotes used in this chapter were taken from the following engineering research articles:

Baker, N. A., Cham, R., Hale, E., Cook, J., & Redfern, M. S. (2007). Digit kinematics during typing with standard and ergonomic keyboard configurations. *International Journal of Industrial Ergonomics* 37: 345–355.

Cheng, Z. Y., & Leong, E. C. (2014). Effect of confining pressure and degree of saturation on damping ratios of sand. In N. Khalili, A. R. Russell, & A. Khoshghalb (Eds.) *Unsaturated Soils: Research & Applications* (Vol. 1, pp. 277–282). Proceedings of the 6th International Conference on Unsaturated Soils, UNSAT 2014, Sydney, Australia, 2–4 July 2014. CRC Press: London.

Han, J., Qiu, W., Hu, J., & Gao, W. (2012). Chemisorption of estrone nylon microfiltration membranes: Adsorption mechanism and potential use for estrone removal from water. *Water Research* 46: 873–881.

Haouala, R., Hawala, S., El-Ayeb, A., Khanfir, R., &, Boughanmi, N. (2008). Aqueous and organic extracts of *Trigonella foenum-graecum* L. inhibit the mycelia growth of fungi. *Journal of Environmental Sciences* 20: 1453–1457.

Leong, E. C., Widiastuti, S., Lee, C. C., & Rahardjo, H. (2007). Accuracy of suction measurement. *Geotechnique* 57(6): 547–556.

Tsakalakos, L. (2008). Nanostructures for photovoltaics. *Materials Science and Engineering R* 62: 175–189.

Wijaya, M., & Leong, E. C. (2014). Modelling shrinkage behavior of soft soils. In *Proceedings of Soft Soils* (Vol. 2, pp. C6.1–C6.6), Bandung, Indonesia, 20–23 October 2014.

Chapter 14

Writing the materials and methods

Theories without data are like daydreams.

Jonathan Rottenberg
The Depths, 2014

The *Materials and Methods* section or chapter usually follows the *Introduction* and *Literature Review* sections/chapters. However, in journal papers and conference articles, the *Materials and Methods* section usually appears after an *Introduction* section which incorporates the *Literature Review* (see *Chapter 13—Writing the introduction and literature review*). The norm in reports and theses is that *Materials and Methods* are presented as a section or chapter after the *Literature Review*.

COMPONENTS OF MATERIALS AND METHODS

Generally, the *Materials and Methods* section/chapter can be regarded as a chronological sequence of procedures taken to collect empirical data. It is an explanation of the data collection process and consists of three components in chronological order: Component 1 is a description of the experimental set-up, Component 2 gives a description of experimental steps or procedures and Component 3 provides the data analysis procedures. A guide of what to include in the three components is summarised in Writing Guidelines 14.1, based on a synthesis of Bruce's (2008) findings and our analysis of engineering reports and theses.

Component 1 can contain up to four steps, A–D. Step A is optional and is usually achieved by a heading which informs the reader about the general content of the *Materials and Methods* section. Alternatively, this can also be accomplished by a diagram of an experimental set-up or a paragraph summarising a macro-perspective of the experiment. Step B is frequently present and consists of two sub-steps: (1) a description of the materials/apparatus used (see Writing Guidelines 14.2 for examples) and (2) the validity and reliability measures taken (see *Chapter 7—Research methodology and research methods* for a fuller explanation of validity and reliability). Step C refers to

Writing Guidelines 14.1 Components of *Materials and Methods* section/chapter

Component 1: Description of experimental set-up

A. Overview of experimental set-up (as a brief paragraph, diagram or heading)
B. Describe the data collection instruments used, in terms of
   i. Materials/apparatus used
   ii. Validity and reliability measures taken
C. Identify the dependent, independent and controlled variables
D. Describe the participants of the study, in terms of
   i. Location of the sample
   ii. Size of the sample
   iii. Traits of the sample
   iv. Context of the sample
   v. Ethical issues of sampling

Component 2: A description of experimental steps taken

A. Describe the steps in the data collection process
B. Justify the steps in the data collection process, by
   i. Stating the purpose of a specific step taken
   ii. Stating the result of a specific step taken
   iii. Highlighting advantages and/or disadvantages

Component 3: Description of data analysis procedures

A. i. State analysis procedures, ii. Justify the data analysis procedures
B. Preview results

---

experimental variables and control variables of your study. If your study involves human participants, Step D is needed to flesh out the characteristics of your human participants.

Component 2 details the experimental procedure in chronological order. Component 2 consists of two steps, A and B. Step A describes the experimental procedures used to collect data. Step B provides the justification for the experimental procedures. You can justify the procedures by pairing the result or outcome of an experimental step with the experimental step, known as means–results relations (Bruce, 2008). Examples of means–purpose and means–results relations are marked by italicised words in Table 14.1. Sometimes it is necessary to highlight the advantages and disadvantages of the experimental steps chosen (Step Biii of Component 2 in Writing Guidelines 14.1).

Component 3 details how the test results are to be analysed and consists of two steps, A and B. The data analysis procedures are described in Step A and if relevant, justification of the procedures, especially if this is not common knowledge or have not been performed before. Typical test results can be included as a preview of what the readers can expect from the tests (Step B).

Writing Guidelines 14.2 illustrates the components and steps in a content analysis of a sample *Materials and Methods* section.

Writing Guidelines 14.2 Content analysis of a sample *Materials and Methods* section

Abstracted from Cheng and Leong (2012)

2 Experimental Setup and Test Procedures

2.1 Soil Properties

The soil tested is Changi reclamation sand which consists mainly of uniform medium grains with minute amount of sea shells (1Bi)[a]. Prior to the experiment, the sand is sieved through sieve No. 8 (2.36 mm) to remove large sea shells (1Bii). The basic soil properties are tabulated in Table 1 (1Bi). Grain size distribution of the sand is shown in Fig. 1.

Table 1   Basic properties of Changi reclamation sand

Soil classification (ASTM D-2487)	SP
G_s	2.66
$\gamma_{d,min}$ (g/cm³)	1.444
$\gamma_{d,max}$ (g/cm³)	1.728
e_{max}	0.842
e_{min}	0.539

2.2 Wave measurement using ultrasonic platens

The experimental setup is shown in Fig. 2 **(1A)**. The top and bottom platens were GCTS ultrasonic platens **(1Bi)**. Details of the platens can be found in Leong et al. (2004) **(1Bi)**. However, instead of using the GCTS standard set-up, the piezoelectric crystals in the transmitter platen were excited using a function generator and amplifier **(2A)**. The signal at the receiver platen was recorded using a digital oscilloscope **(2A)**. A porous disk and a 1 bar high-air entry ceramic disk were fixed on the top and bottom platens, respectively **(2A)**. The sand was first compacted moist to a relative density of 70% in a metal split mould **(2A)**. The specimen was then transferred to the triaxial cell **(2A)**. A rubber membrane was placed over the specimen **(2A)**. The top platen was then placed on top of the specimen and secured with O-rings **(2A)**. Saturation was performed at an effective confining pressure of 10 kPa *by* applying a confining pressure of 210 kPa and back pressure of 200 kPa *using* a GDS digital pressure volume controller (DPVC) **(2Bii)**. A small suction was applied at the pressure line which was connected to the top platen **(2A)**. During saturation, V_p was recorded **(2A)**. Saturation continued until the following conditions were met: (1) V_p remains constant; (2) no air bubbles were observed in the air pressure line; (3) increase in water level in the flushing pot; and (4) Skempton's pore pressure parameter, B > 0.95 **(2A)**.

Upon saturation, the top platen's pressure line was connected to an air pressure system at 250 kPa **(2A)**. The specimen was then consolidated at a net normal stress (σ-u_a) of 10 kPa and the amount of water that drained out was monitored *using* the DPVC **(2Bii)**. Matric suction was applied *by* reducing the pore-water pressure (u_w) while maintaining the pore-air pressure (u_a) at 250 kPa **(2Bii)**. Values of V_p and V_s were measured at equilibrium matric suctions of 0, 0.5, 1.0, 1.5, 2.0, 3.0 and 5.0 kPa **(2A)**. At each matric suction, V_p and V_s were obtained *by* exciting the ultrasonic platen at frequencies of 10, 20, 50, 100 and 200 kHz **(2Bii)**. Upon completion, the whole specimen was dried in the oven *for* water content determination **(2Bi)**. The water contents at various matric suctions were back-calculated from the final water content and the soil-water characteristic curve (SWCC) of the sand is shown in Fig. 3 **(3Ai)**.

[a] Bracketed term at the end of each sentence indicates a step, for example **(1A)** indicates step A of Component 1, **(2A)** indicates step A of Component 2 and so on as shown in Writing Guidelines 14.1.

Note: Words in italics indicate means-purpose and means-results relations.

Table 14.1 Means–purpose, means–result relations

	Description	Examples
Means–result	Involves a statement of how a particular result is/was/will be achieved	Argon was purified by passing it through a series of columns (order Result Means)
Means–purpose	Involves how an action is/was/will be undertaken with the intention of achieving a particular result	Aliquots from the stock were used for preparing the working solutions (order Means Purpose)

Source: Bruce, I., Journal of English for Specific Purposes, 7, 45, 2008, Copyright (2008) by Elsevier.

SOME CONSIDERATIONS IN WRITING MATERIALS AND METHODS

Your methodological design should be based on sound empirical concepts of validity and reliability (see *Chapter 7—Research methodology and research methods*). Do not describe experimental procedures that can be found in standards, for example local standards, international standards (ISO), American Society for Testing and Materials (ASTM) standards or British standards (BS). Instead cite the reference standard and test number, for example 'The tests were conducted in accordance to ASTM D5035-11 (2011)'. The words 'in accordance to' indicate compliance to the test standard, ASTM D5035-11 (2011), and readers should refer to ASTM D5035-11 (2011) for the detailed experimental procedures and data analysis. For experiments/tests that are new or devised in your research, use methodological triangulation (citing different sources where similar experiments/tests were performed) to justify your choice of experimental procedures.

The *Materials and Methods* section should be concisely written. The experimental procedures should be supplemented with diagrams, schematic drawings or figures as far as possible (see *Chapter 21—How to create figures*). Good illustrations of how the *Materials and Methods* section should be written are given in the standards.

SOME KEY LANGUAGE FEATURES IN MATERIALS AND METHODS

Verb tenses

Similar to the context-sensitive verb tense used in the introduction, the tense varies according to the content of the *Materials and Methods* section. The simple past tense and/or past perfect tense are used to describe any step taken to collect data, especially when the *Materials and Methods* section is a recount or summary of the data collection process. The past perfect tense is extensively used to indicate the chronological precedence of an experimental step taken before succeeding steps. The simple present tense or present perfect tense is used to refer to advantages or disadvantages of an experimental procedure or method as claimed by previous studies. See *Chapter 10— Grammar, punctuation and word usage guide* for an explanation of tenses.

Passive voice

The passive voice is the predominant grammatical construction in *Materials and Methods* sections. The passive voice is regarded as an objective language device which emphasises the product or process rather than the experimenter. See *Chapter 10—Grammar, punctuation and word usage guide* for an explanation of passive voice.

CHECKLIST FOR MATERIALS AND METHODS

Does your *Materials and Methods* include

☐ A justification of the methodological approach and research design if both of them are new or controversial?
☐ A description of the data collection procedures?
☐ An explanation of the data analysis procedure?

INTERESTING FACTS

One of the most outstanding theses to be written in the twentieth century is Claude Shannon's 1937 master's degree thesis called 'A symbolic analysis of relay and switching circuits' submitted at the Massachusetts Institute of Technology. In his thesis, Claude explained how electrical applications of Boolean algebra could form and address any logical and numerical relationship. This forms the foundation of computers we know today.

REFERENCES

Bruce, I. (2008). Cognitive genre structures in Methods sections of research articles: A corpus study. *Journal of English for Academic Purposes* 7: 38–54.
Cheng, Z. Y., & Leong, E. C. (2012). Ultrasonic testing of unsaturated soils. In L. Laloui & A. Ferrari (Eds.), *Multiphysical Testing of Soils and Shales* (pp. 105–110). Luasanne, Switzerland: Springer.

Chapter 15

Writing the results and discussion

> Huge volumes of data may be compelling at first glance, but without an interpretive structure they are meaningless.
>
> Tom Boellstorff
> *Ethnography and Virtual Worlds: A Handbook of Method, 2012*

Results and Discussions are the main thrust of the research report or thesis—they present the findings of the study and the author's discussion or interpretation of these findings. In the *Results and Discussion* section or chapter, you present results that answer your research questions or address your objectives or hypotheses. You may need to translate the results from figures in a table, figure or equation to descriptions or statements about their significance or relationships in your study's context. Also, you may need to refer to information related to your methodology and/or the technical/theoretical background of your methods and materials.

COMPONENTS OF RESULTS AND DISCUSSION

A comprehensive breakdown of the components of *Results and Discussion* is shown in Writing Guidelines 15.1 based on a synthesis of findings by Swales and Feak (2012), Peacock (2002) and our analysis of engineering reports and theses.

Component 1 includes three steps, namely, a statement of background information (Step A), a restatement of methodological information (Step B) and a restatement of research questions/hypotheses (Step C). According to Swales (1990), the indication of background information (Step A) '...is employed by authors when they wish to strengthen their discussion by recapitulating main points, by highlighting theoretical information, or by reminding the reader of technical information'. The restatement of methodological-related information (Step B) is specific to a result which the author presents in writing. It is typical in some engineering texts to contextualise findings with background or methodological information even though relevant methodological information is already

Writing Guideline 15.1 Components of *Results and Discussion*

Results

Component 1: Present meta-textual information

A. State background information

B. Restate methodological information

C. Restate research questions/hypotheses

Component 2: Present finding

A. Direct readers' attention to table or figure

B. Highlight finding

Discussion

Component 3: Interpret findings

A. Explanation of finding (suggest reasons; argue for a cause–effect relationship)

B. Make a generalisation based on finding

C. Compare with previous findings (and show consistency or inconsistency, with explanation)

D. Support from theories

E. Make an inference (with sufficient support from findings and relevant literature)

presented in the *Materials and Methods* section. Step C is a restatement of the research question/hypothesis related to a finding. This step is often found in undergraduate and postgraduate reports or theses but usually not found in research articles.

Step C in Component 1 is followed by Component 2, presentation of findings. You should present only findings that address your research questions/hypotheses. If your data analyses reveal surprising findings not related to your research questions/hypotheses, you may want to add research questions to your *Introduction*. However, you should first discuss the ideas with your supervisor. The presentation of findings consists of the following steps. Step A is necessary to direct readers' attention if there are tables or figures which need to be referred to in the text. For example, you can use phrases such as 'with reference to Table 2', 'see Figure 5.3' or simply state the figure or table number in parenthesis, such as '(Table 3.4)' at the end of a relevant sentence. Step B is required to highlight an important finding or result, with technical or mathematical explanation. This explanation is limited to empirical observations. Explanations of causality (relationship of cause and effect) are indicated in Step A of Component 3 in the typical content structure.

Component 3 is the start of the discussion section/chapter which details your comments or interpretations of your results. There are five types of comments (A–E) to include as discussion of findings. Type A comments give or suggest reasons or argue for a causality which can be indicated by words such as *cause, attributed to, contributed and led to.* Type B comments make a generalisation or deduction based on findings, which can be indicated by words such as *generally, overall, typical of and usually.* Type C comments compare finding/s with previous findings to show consistency or inconsistency, which are marked with in-text citations (see *Chapter 18— Referencing*) and can be indicated by words such as *in agreement with, dissimilar to and corroborate with.* Type D comments state explanations of findings based on existing theories, which are marked with in-text citations if theories are found in scholarly works and can be indicated by words such as, *agree with, supported by, consistent with and substantiated by.* Regarding Types C and D comments, and the cited studies should already be discussed in the *Literature Review* so that you can refer to them in your discussion of results. Type E comments contain inferences which could include explanations or generalisations that are logically sound but lack empirical support. Inferences can be indicated by hedges (words that expresses uncertainty) such as *probably, hypothetically, subject to further empirical investigations and a possibility.*

Note: Not all the various types of comments need to be present in a *Results and Discussion* section.

A sample content analysis of a sample *Results and Discussion* section is shown in Writing Guidelines 15.2.

Writing Guidelines 15.2 Content analysis of a sample *Results and Discussion* section

Abstracted from Wijaya and Leong (2014)

Comparison of shrinkage curve equations for soft soils

Six soft soils (NSF clay, kaolin clay, Kasaoka clay, Kurita clay and two marine clays subjected to different loading stages) will be evaluated in this paper using three different shrinkage curve equations which are Braudeau et al. (1999) equation, Fredlund et al. (2002) equation and Leong and Wijaya (2015) equation **(1B)**[a].

All of the soft soils investigated in this paper only have two linear segments which is the most common type of shrinkage curve **(1A)**. Therefore, the three shrinkage curve equations which are able to model more than two linear segments shrinkage curve were simplified to model two linear segments shrinkage curve only **(1B)**. The NSF clay and kaolin are intended for industrial use, Kasaoka clay was made from crushed mudstone taken from an area near Kasaoka city while Kurita clay is a natural soil sampled from Nagano city (Umezaki and Kawamura 2013) **(1A)**. The two marine clay soils were from Termunten on the northern coastal area of the Netherlands (Kim et al. 1992) **(1A)**. The two marine clays were under different loading stages and therefore have different initial gravimetric water contents **(1A)**.

The Braudeau et al. (1999) equation (BEA model) was shown to be versatile in modelling a shrinkage curve as it uses explicit parameters (water content and void ratio at the point of convergence between each linear segments) and has high accuracy (Cornelis et al. 2006a; Leong and Wijaya 2015) **(1A)**. The equation divides the shrinkage curve into linear parts and a non-linear part **(1A)**. The linear parts are used to represent the zero shrinkage line and loading line while the non-linear part is used to represent the residual shrinkage phase **(1A)**. It is expected that the void ratio at the shrinkage limit e_{SL} is the same as the minimum void ratio e_{min} as the shrinkage curve has reached the zero shrinkage phase and

(Continued)

further reduction in water content will not cause additional shrinkage **(1A)**. Braudeau et al. (1999) equation for a two linear segment shrinkage curve is given as:

$$e = \begin{cases} e_{min}; \ w \leq SL \\[2mm] e_{SL} + (e_{AE} - e_{SL}) \dfrac{f}{\exp(1) - 2}; \ SL < w \leq w_{AE} \\[2mm] e_s + (e_{AE} - e_s) \dfrac{w - w_{AE}}{w_s - w_{AE}}; \ w > w_{AE} \end{cases} \tag{1}$$

and,

$$f = \exp\left(\frac{w - SL}{w_{AE} - SL} \right) - \left(\frac{w - w_{SL}}{w_{AE} - SL} \right) - 1 \tag{2}$$

where e is the void ratio, e_{AE} is the void ratio at the air entry point and e_s is the void ratio at full saturation **(1A)**.

Fredlund et al. (2002) equation (FEA model) was proposed for use in geotechnical engineering to describe shrinkage curves with two linear segments **(1A)**. However, it has not been compared with other shrinkage curve equations **(1A)**. The equation is given as:

$$e(w) = e_{min} \left(\frac{w^C}{SL^{'C}} + 1 \right)^{1/C} \tag{3}$$

where SL' is the apparent shrinkage limit which corresponds to the water content at the intersection of the loading line and the zero shrinkage line and C is a model parameter which requires curve fitting to obtain its value **(1A)**. Typical C values for different types of soils are given in Fredlund et al. (2002) **(1A)**.

Leong and Wijaya (2015) equation (LW model) was proposed to provide a single continuous equation with explicit parameters to model a shrinkage curve with more than two linear segments **(1A)**. The simplified form of Leong and Wijaya (2015) equation for a two linear segment shrinkage curve is given as:

$$e(w) = e_{min} + \frac{G_S}{2S_0} \left\langle w + \frac{1}{k} \ln\left\{ \frac{\cosh\left[k(w - SL')\right]}{\cosh(k..SL')} \right\} \right\rangle \tag{4}$$

where G_S is the specific gravity, S_0 is the initial degree of saturation and k is a parameter which is related to the curvature **(1A)**. Parameter k can be obtained from SL and AE by the following relationship:

$$k = \frac{2}{AE - SL} \tag{5}$$

For the above-mentioned three shrinkage curve equations, only the C parameter in the FEA model needs to be curve-fitted (implicit-parameter) while the BEA model and LW models use explicit parameters **(1A)**. Therefore, in the comparison the C parameter in the FEA model was curve-fitted while the other parameters in the FEA model used were the same parameters as those used in the BEA and the LW models **(1B)**. The results are shown in Figures 2 and 3 with the coefficients of determination, R^2. The R^2 for each model is indicated by the subscript **(2A)**. The results show that the three equations give very high R^2 values **(2B)**. Overall, the LW model performed the best,

(Continued)

Writing Guidelines 15.2 (Continued)   Content analysis of a sample *Results and Discussion* section

followed by the FEA model and then the BEA model **(3B)**. However, in Figure 3b, the FEA model gives a shrinkage curve that lies on the left of the loading line although it has a high R^2 value **(2B)**. Another observation in Figure 3b concerns the BEA model **(2A)**. The shrinkage curve for the BEA model crosses the loading line which is theoretically not plausible **(2B)**. This is because the non-linear part of the BEA model does not consider the existence of the loading line **(3A)**.

[a] Bracketed term at the end of each sentence indicates a step, for example, **(1B)** indicates step B of Component 1, **(2B)** indicates step B of Component 2 and so on as shown in Writing Guidelines 15.1.

ORGANISING RESULTS AND DISCUSSION

There are two ways of organising your *Results and Discussion*. You can present all your findings together and then follow with a discussion of the findings either within the same section or chapter or in separate sections or chapters. This presentation format is called the sequential pattern. This pattern can be represented as follows:

Sequential pattern: $R1 + R2 + R3 + R4 + D$

[R = Results; D = Discussion]

Of course, this pattern can be extended if you have more than four findings or results to report.

The second way of ordering your results and discussions is to present a finding and then discuss or comment on it before presenting the next finding followed by discussion. This presentation format is called the alternating pattern and can be represented as follows:

Alternating pattern: $R1 + D + R2 + D + R3 + D + R4 + D$

[R = Results; D = Discussion]

The alternating pattern is best if you have many different results and you have specific discussions on each of these results. The sequential pattern is better when you have several different results to which one discussion can apply.

COMMON LOGICAL PITFALLS IN RESULTS AND DISCUSSION

Some common problems that may occur in the writing of *Results and Discussion* sections, particularly when you comment on a finding or findings, are as follows:

False analogy—Divergent items are compared and equated as evidence. Example: Selling synthetic meat* to people is like selling poisoned food to them. They will be poisoned, fall ill and die.

---

* Synthetic meat is also known as test-tube meat, in vitro meat, victimless meat, cultured meat, tubesteak, cruelty-free meat, shmeat and artificial meat.

Post hoc or false cause ('after this, therefore because of this')—When an event follows another, it is assumed that the former caused the latter. Example: An increase in artificial meat sales last year was followed by an increase in heart disease, atherosclerosis and asthmatic cases this year. Therefore, we should ban artificial meat because they cause heart disease.

Slippery slope—Slippery slope is a prediction of a drastic snowballing effect without substantial evidence. Example: The unabated use of artificial meat will lead to a scarcity of natural food, followed by the collapse of farms globally and then a world economic depression.

False dilemma ('either/or')—It is insisted falsely that one or another of two alternatives must be chosen though there may be many other positions besides the two. Example: We are faced with a grave situation—either we ban artificial meat and eliminate health problems or we promote artificial meat and be plagued with skyrocketing health problems.

SOME KEY LANGUAGE FEATURES OF RESULTS AND DISCUSSION

Verb tenses

Similar to the variation in verb tense in specific content components found in the *Introduction and Materials and Methods* sections, the verb tense varies in the *Results and Discussion* section. The simple past tense is used for Part 2 Component 1 (restating research questions) and Part 2 Component 2 (referring to tables or figures). In this case, tense is used interactively with the reader to reiterate the present study's research questions/hypotheses and direct or guide their attention. The simple past and past perfect tense are used to highlight findings (Part 2 Component 3). The tense is used for its temporal sense, to signal that data collection and analysis were conducted.

The simple present tense or present perfect tense is used in all the components of Part 3. Tense is used rhetorically to underlie the currency of the author's arguments and imply the author's support for the claims. See *Chapter 10—Grammar, punctuation and word usage guide* for an explanation of tenses.

Voice

Voice is a language-based impression of your identity as an author in your writing. In other words, voice refers to language features that highlight or downplay references to you as an author or other authors. Objective or impersonal language downplays your authorial identity. It should be recognised that your writing is subjective because it is written from your

Table 15.1 Impersonal language features

Impersonal language features	Examples
Passive voice	The results *were culled.*
Nominalisation	The *culling* of results...
Absence of self-mentions/author pronouns/ possessive adjectives	
'It is...that' clauses	*It is* established *that* culling of data...
'There is + noun' fronted sentences	*There is an assumption* that...
Attributive tags	*According to* Ong (2011),...
Personification or associations of inanimate things with verbs usually used with people	*The results demonstrate* that...

Table 15.2 Personal language features

Personal language features	Examples
Self-mentions/author pronouns/possessive adjectives	I; we; my; our; us
Engagement markers	As you can see; you will have noted that; consider whether

perspective. However, you can increase the acceptability of your subjective ideas by presenting them in impersonal language to imply that your ideas are largely unbiased. Most engineering writing employs objective language. Some impersonal language features and examples are detailed in Table 15.1.

Infrequently, engineering writing may involve subjective or personal language which highlights your authorial identity. There are two key subjective language features you can use. You may use self-mentions judiciously if you want to indicate ownership of your ideas, to imply your authority over a knowledge claim, or to hedge your claims as a personal opinion or inference. Also, you may want to use engagement markers to bring readers into the text as participants to highlight your anticipations of their objections or views, to direct them along a line of reasoning or to assume their solidarity with your views. Some personal language features and examples are detailed in Table 15.2.

Evaluative language

Evaluative language conveys personal attitudes, emotions, assessments and propositions from you as an author and/or other authors (Gray & Biber, 2012). According to Thompson and Hunston (2000), evaluative language or stance can express:

1. Positive–negative evaluations
2. Certainty–uncertainty in relation to a claim

Table 15.3 Evaluative language features and examples

Strategies	Purpose	Examples
Hedges	Indicate uncertainty—to withhold your full commitment to a statement	May; might; possibly; seemed to; appeared to
Boosters	Indicate certainty—to reveal your certainty about a statement	Clearly; definitely; without doubt
Attitude markers	Indicate positive or negative evaluations, or expectedness/ unexpectedness, or importance/ unimportance	Interestingly, surprisingly, unfortunately, importantly, significantly, scholarly, sound, problematic

3. Expectedness–unexpectedness of a claim
4. Importance–unimportance of a claim

Some key evaluative language features and examples are detailed in Table 15.3.

TIPS ON TYPING EQUATIONS

It is very likely that you will be typing equations in the *Results and Discussion* section. Equations are also commonly found in the *Literature Review* and *Materials and Methods* sections of engineering reports and theses. Many students are unaware that there is an equation editor in Microsoft Office suite. It is quite likely that you will need to type equations in Microsoft Office Word and PowerPoint. Word 97–2003 and Word 2007, 2010 have different methods for you to type an equation.

To type an equation such as $x = \dfrac{-b \pm \sqrt{b^2 - 4ac}}{2a}$ in Microsoft Office Word 97–2003:

1. Select the Insert tab and then select Object in the Text group. Click the dropdown box and select Object… to bring up the Object dialog box (Figure 15.1).
2. Look for Microsoft Equation Editor 3.0 under the Create New tab. Select it by clicking it and click the OK button (Figure 15.1). This will bring up the Equation toolbar (Figure 15.2) and an equation text box on your Word document for you to start typing the equation.
3. Type $x = \dfrac{-b \pm \sqrt{b^2 - 4ac}}{2a}$ as per normal except that for the math symbols, you type by clicking on the appropriate symbol in the Equation tool.
4. Click outside the equation text box when you are done.

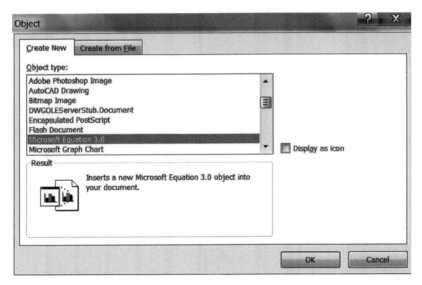

Figure 15.1 Object dialog box under *Insert, Text group, Object* of Microsoft Office Word 97–2003.

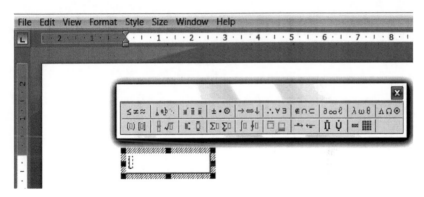

Figure 15.2 Microsoft Equation 3.0 toolbar and equation text box in Microsoft Office Word 97–2003.

Microsoft Office 2007 and 2010 provide a more user-friendly equation tool for you to type equations. The Microsoft Equation 3.0 editor in Word 97–2003 is an add-in which may still be available in your version of Word 2007 or 2010 if you had installed it. To type $x = \dfrac{-b \pm \sqrt{b^2 - 4ac}}{2a}$ by using the newer Equation editor in Word 2007 and 2010:

1. Select the Insert tab and you will see a Symbols group as shown in Figure 15.3.
2. Select the Equation dropdown box and you will see groups of Built-In equations.

3. Select the available Built-in quadratic equation $x = \dfrac{-b \pm \sqrt{b^2 - 4ac}}{2a}$ as shown in Figure 15.4 and you are done.

If you are typing other forms of equations, you can either select a Built-In equation that is close to the form of your equation and modify it by selecting the appropriate math symbol in the Equation Design menu bar as shown in Figure 15.4, or you can select Insert New Equation and type in the equation in the Equation text box.

A word of caution: If you save a Word 2007 or 2010 document in the earlier version Word 97–2003, the equation will be saved as an image and you cannot edit the equation anymore. Similarly if you open a Word 97–2003 document in Word 2007 or 2010, you can only edit the equation if the Microsoft Equation 3.0 add-in has been installed.

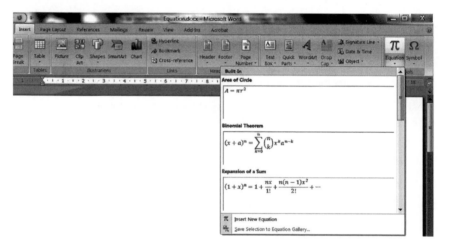

Figure 15.3 *Equation* dropdown menu in the *Symbols* group of Microsoft Office Word 2007.

Figure 15.4 *Equation Design* menu in Microsoft Office Word 2007, 2010.

CHECKLIST FOR RESULTS AND DISCUSSION

Does your results and discussion include
☐ Presentation of results that are related to your research questions/hypotheses?
☐ Technical or mathematical explanation of your findings (without argument for causality)?
☐ References to methodological and background information?
☐ Explanation (or argument for causality) or comparison with prior findings, or evaluation of your findings in light of theories?

INTERESTING FACTS

Richard Feynman, the renowned theoretical physicist in path integral formulation of quantum mechanics and Nobel laureate, completed his PhD thesis at Princeton University in 1942 titled 'The Principle of Least Action in Quantum Mechanics'. His thesis laid the groundwork of path integral technique and Feynman diagrams.

REFERENCES

Gray, B., & Biber, D. (2012). Current conceptions of stance. In K. Hyland & C. S. Guinda (Eds.), *Stance and Voice in Written Academic Genres*. Basingstoke: Palgrave Macmillan.
Peacock, M. (2002). Communicative moves in the discussion section of research articles. *System* 30: 479–497.
Swales, J. (1990). *Genre Analysis: English in Academic and Research Settings*. Cambridge: Cambridge University Press.
Swales, J., & Feak, C. (2012). *Academic Writing for Graduate Students* (3rd ed.). Ann Arbor: University of Michigan Press.
Thompson, G., & Hunston, S. (2000). Evaluation: An introduction. In S. Hunston & G. Thompson (Eds.). *Evaluation in Text*. Oxford: Oxford University Press.
Wijaya, M., & Leong, E. C. (2014). Modelling shrinkage behaviour of soft soils. In *Proceedings of Soft Soils*, Bandung, Indonesia, 20–23 October 2014, Vol. 2, pp. C6.1–C6.6.

Chapter 16

Writing the conclusion

(I)n order to refute a conclusion, you have to put forth the best possible argument for it

Rebecca Newberger Goldstein
Plato at the Googleplex: Why Philosophy Won't Go Away, 2014

The *Conclusion* (or *Conclusions*) is the last major section of your report or thesis. This section is sometimes called *Discussion, Discussion and Conclusion(s)*, or *Summary and Conclusions*. Instead of writing a *Conclusion* section, some researchers incorporate it into the *Results and Discussion* section in journal papers (Bunton, 2005, p. 212). Whether you use the term *Conclusion* or another term is really a matter of individual preference and style. Whichever term is used for this final section or chapter, the writing conventions reflect common features.

The general function of the *Conclusion* is to bring closure to the research questions stated in the *Introduction*. In the *Conclusion* the writer essentially contrasts pre-contribution with post-contribution so that 'what was unproven, unverified, unexplained, unknown, partial or limited is now proven, verified, explained, known, complete or general' (Lebrun, 2007, p. 199).

FUNCTIONS OF A CONCLUSION SECTION

The *Conclusion* section typically performs the following functions (Bunton, 2005, p. 213):

1. To remind the reader of the aims of your study (e.g. research questions/hypotheses) and key methodological features of your study
2. To summarise the most important findings and conclusions of your study
3. To evaluate the importance and significance of your study with commentary on its contribution to the development of theory and research

4. To point out the practical applications of your findings
5. To point out any limitations (if any) of your study
6. To recommend areas for further research

COMPONENTS OF THE CONCLUSION SECTION

Weisberg and Buker (1990) has pointed out that the *Conclusion* section in research reports typically has four basic components, shown in Table 16.1. They perform the functions of the *Conclusion* section identified earlier in the chapter.

ILLUSTRATION OF A CONCLUSION SECTION IN A RESEARCH PAPER

Writing Guidelines 16.1 illustrates how the components are reflected in a research paper. The language features used to perform the various functions are highlighted.

LANGUAGE FEATURES OF THE CONCLUSION

In this part of the chapter, we examine the language conventions used to present the information in the different components of the *Conclusion*. The language features discussed here are verb tenses and modal verbs.

Table 16.1 Components of *Conclusion*

Components	Steps
1. Restatement of aims and methodological approach of study	A. Restate aims of study (research questions/hypotheses) B. Restate key features of research methodology & methods
2. Summary of findings/conclusions	A. Review key findings B. Explain or compare
3. Evaluation of study's contribution	A. Point out significance of findings (for theory & research development and for practical applications) B. Point out practical applications of your study C. Identify limitations (if any)
4. Recommendations for future research	A. Recommend areas for further research

Writing Guidelines 16.1 Content analysis of a sample *Conclusion*

Component 1: Restatement of aims	A. Restate research aim/focus	*In this study*, the use of photogrammetry to measure volume change of a triaxial specimen was evaluated.
Component 2: Summary of findings	A. Summarises key findings	*Key findings* from the current study include the need for (1) redundancy of images (to ensure greater overlap of the images), (2) texturing of rubber membrane (to ensure better 'stitching' of the images), (3) high camera resolution (to obtain better image quality), (4) allowance for operator's error (in obtaining the images), and (5) allowance for compliance error of the digital pressure volume controller.
Component 3: Evaluation of study's contribution	A. Indicate significance of study (for research development)	An *innovative set-up* was designed to allow the camera to rotate and take all-around photos of the specimen at a constant height and distance. The applicability of this technique was investigated through two separate consolidation tests on kaolin and Changi sand. By comparing the image processing results to DPVC readings, a quantitative evaluation was generated.
	B. Point out practical applications & make recommendations	*The applicability of this technique to measure volume change* was investigated through two separate consolidation tests on kaolin and Changi sand and comparing the volume estimation of image processing results with the DPVC readings. The comparison shows that photogrammetry *can be employed for unsaturated triaxial soil tests if the volume change is greater than +/−100 mm³*. This translates to volumetric strains greater than +/−0.12, 0.05%, 0.02% and 0.01% for specimens of diameter 38, 50, 70 and 100 mm, respectively, and height to diameter ratio of 2.
	C. Identify limitations of study	However, several aspects with regard to the existing study and set-up *could be further improved* such as the use of soft, homogenous lighting as well as professional-grade cameras. Non-isotropic consolidation and loading on a soil specimen could be examined. Automation of both the rotation and image capturing process by the camera could be developed to reduce operator's error and increase efficiency in positioning of the camera.
Component 4: Recommendation for future research	A. Recommend area for further research	Compliance of the DPVC *should be further investigated* as it showed a very different volume change at low volume (and low pressure).

Source: Lim, B. J. M., & Leong, E. C. *Unsaturated Soils: Research & Applications, 2014.*

Verb tenses

As Weisberg and Buker (1990) observed, verb tenses in the *Conclusion* vary depending on the type of information being presented (p. 170). The verb tense most commonly used in the first two components when referring to purpose, hypothesis and findings is the *simple past tense*. In the third component when commenting on the significance and limitations of the study, the simple present tense is usually used. When indicating areas for future research in the final component, the simple present tense and modal verbs are normally used (see Table 16.2).

Modal verbs

This section deals with a complex language area which is important in the *Conclusion* (and also the *Results and Discussion*) section: modal verbs. Glasman-Deal (2010, pp. 150–166) identified modal verbs that are commonly used in scientific and technical writing and listed as most common are *may, might, could, can, should, ought to, need to* and *must*.

Modal verbs are used to indicate the degree of certainty with which a statement is made or to modify the 'truth value' of a statement. Compare the two sentences below:

1. The drop in acceleration pressure *was* due to density change in the fluid.
2. The drop in acceleration pressure *may have been* due to density change in the fluid.

Table 16.2 Correspondence between verb tenses and functions

Function	Verb tense	Example
Restating aim *(Component 1A)*	Simple past	This research *attempted* to assess two methods for air purification.
Restating hypothesis *(Component 1A)*	Simple past	It *was* originally *assumed* that …
Summarise findings *(Component 2A)*	Simple past	It *was found* that … The test *showed* that …
Explain/compare findings *(Component 2B)*	Simple present	The results *are* consistent with … The findings *differ* from those of …
Evaluate study's contributions *(Component 3A & B)*	Simple present	This research *adds* to the body of knowledge … These findings *provide* evidence/*lend* support to the assumption that…
Identify limitations of study *(Component 3C)*	Simple present	The small sample size *is* a limitation of the research.
Recommend future research *(Component 4A)*	Simple present	The small sample size *is* an opportunity for future research.

There is no modal verb in sentence 1 as the writer is stating with *absolute certainty* that the drop in pressure was caused by the density change in the fluid. In sentence 2, however, the modal verb *may* indicates that the writer considers the density change in the fluid as a *possible* cause for the drop in pressure.

Modal verbs are commonly used when making recommendations for practical applications and for further research. In the excerpt below taken from a journal paper, design suggestions and recommendations are presented with the modal *should*.

> Architects, engineers, and landscape architects *should take into account* the safety of installers and future maintenance workers in the design of the built environment and rooftop vegetation. Green-building certification organisations *should use* these design suggestions and include the safe design of vegetated roofs as a precondition for credit allowance.
>
> Behm, M. (2010). Safe design suggestions for vegetated roofs. *Journal of advanced concrete technology* 8(2): 258.

DO'S AND DON'TS IN WRITING THE CONCLUSION SECTION

Table 16.3 identifies the common mistakes (the *Don'ts*) in writing the *Conclusion* section and how to avoid them (*the Do's*).

Table 16.3 Do's and don'ts of writing the *Conclusion*

Do's	Don'ts
Include only findings presented previously in the document.	Do not include new findings in this section.
Base each conclusion soundly on material/ evidence previously stated in the document.	Do not neglect to ensure that each conclusion is related to specific material presented previously.
Make specific statements.	Avoid vague and generalised statements.
Interpret results or observations.	Do not merely repeat findings from the *Results and Discussion* section without interpretation.
Be modest in stating the significance of your study.	Do not exaggerate the significance of your findings.
State the limitations of your research and recommend areas of future research.	Do not treat your research findings as the final word on the topic.
Make sure the contents in the *Introduction* match those in the *Conclusion*	Do not neglect to make sure that the *Introduction* and *Conclusion* are logically linked.

The *Conclusion* section of the report or thesis is mainly concerned with reporting and interpreting the results and conclusions of your study in relation to the objectives and scope stated in the *Introduction* section. As in the *Introduction*, researchers in the *Conclusion* section indicate where and how their work fits into the research 'map' of their field. As Weisberg and Buker (1990, p. 160) points out, in the *Conclusion* 'you step back and take a broad look at your findings and your study as a whole'.

Finally, remember to check that the contents of the *Conclusion* match those in the *Introduction*. The *Conclusion* should address the objectives and scope stated in the *Introduction*. Read the *Conclusion* and the *Introduction* together to check that there is no contradiction.

CHECKLIST FOR WRITING THE CONCLUSION

☐ Is the *Conclusion* chapter well structured?
☐ Is there a logical link between the *Introduction* and the *Conclusion* chapters?
☐ Have you stated your *most important* results?
☐ Have you given an interpretation of these results, rather than just restating them?
☐ Is each conclusion drawn based on evidence presented previously in the report?
☐ Have you pointed out the importance, significance, and contribution of your work?
☐ Have you indicated the limitations (if any) of your work?
☐ Have you avoided stating limitations of your work in negative language?
☐ Have you indicated areas for future research that are the logical extension of your work?
☐ If appropriate, have you included possible practical applications of your findings?

INTERESTING FACTS

After the *Abstract*, the *Introduction* and *Conclusion* are the most important chapters in a report or thesis. According to Brabazon (2010), short introductions indicate 'start of deeper problem' and short conclusions indicate that students are unable to grasp the significance of their research or that they are too tired to write a good conclusion.

The length of the *Conclusion* section depends on the length of the report or thesis. The length of reports and theses depends on a number of factors: field of research, research method, research topic and sometimes page limit imposed by your institution. Marcus W. Beck plotted the length of master's theses and PhD dissertations from the University of Minnesota as shown in Figure 16.1. The master's thesis data are from 2009 to 2014 and contain 930 records. The PhD dissertation data are from 2006 to 2014 and contain 3,037 records. Figure 16.1 shows at the University of Minnesota, the typical lengths of master's theses and PhD dissertations range from 50 to 150 pages and from 50 to 350 pages, respectively. A reasonable guide on the length of the various sections of a report or thesis in comparison to the whole is: *Introduction* (10%–20%), *Literature Review* (10%–20%), *Materials and Methods* (10%–20%), *Results and Discussion* (20%–40%) and Conclusion (5%–10%). Using this guide, the conclusion for a 50-page report or thesis should be about 2–5 pages, for a 150-page report or thesis it should be about 7–15 pages and for a 350-page report or thesis it should be 17–35 pages.

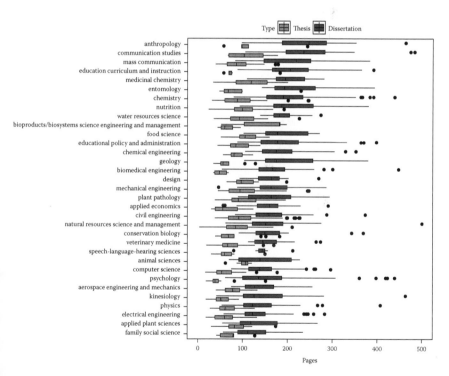

Figure 16.1   Comparison of dissertation and thesis page lengths for majors having both degree programs in the University of Minnesota database (boxes represent the median, 25th and 75th percentiles, the whiskers represent 1.5 times the interquartile range, and outliers are beyond the whiskers). (From https://beckmw.wordpress.com/2014/07/15/average-dissertation-and-thesislength-take-two/. Reproduced with permission from Marcus W. Beck.)

REFERENCES

Brabazon, T. (2010). *How not to write a PhD thesis*. Times Higher Education. Retrieved from http://www.timeshighereducation.co.uk/news/how-not-to-write-a-phd-thesis/410208.article.

Bunton, D. (2005). The structure of PhD conclusion chapters. *Journal of English for Academic Purposes* 4: 207–224. doi:10.1016/j.jeap.2005.03.004.

Glasman-Deal, H. (2010). *Science Research Writing for Non-native Speakers of English*. London: Imperial College Press.

Lebrun, J. L. (2007). *Scientific Writing: A Reader and Writer Guide*. Hackensack, NJ: World Scientific.

Lim, B. J. M., & Leong, E. C. (2014). Photogrammetry for volume change measurement in unsaturated soil tests. In N. Khalili, A. R. Russell, & A. Khoshghalb (Eds.). *Unsaturated Soils: Research & Applications* (Vol. 2, pp. 1605–1610). London: Taylor & Francis Group. Proceedings of the 6th International Conference on Unsaturated Soils, UNSAT 2014, 2–4 July 2014, Sydney, Australia.

Weisberg, R., & Buker, S. (1990). *Writing Up Research: Experimental Research Report Writing for Students of English*. Englewood Cliffs, NJ: Prentice Hall Regents.

Chapter 17

Writing the abstract
and front matter

Science is fundamentally about communication. Un-communicated
science in essence does not exist.

Simeon J. Yates, Noel Williams and Ann-Florence Dujardin

*Writing Geology: Key Communication Competencies
for Geoscience, 2005*

The *Abstract*, sometimes called a summary, is a succinct synopsis of the
report or thesis usually consisting of less than 300 words. It is the first part
of the report or thesis that the reader will read. Conversely, it is the last part
of the report or thesis to be written because the *Abstract* is based on the
content of the entire report or thesis (see *Chapter 12—Strategies for writing
a good report or thesis*). The abstract contains only key information found
in the report or thesis. In some conferences, only an extended abstract (a
longer version of the abstract) is published. In essence, an abstract is a mini-
article (Gillaerts & de Velde, 2010).

COMPONENTS OF AN ABSTRACT

The *Abstract* is referred to as a 'mini-article' as its format corresponds to that
of the report or thesis in consisting of four main components: *Introduction,
Materials and Methods, Results and Discussions,* and *Conclusion* (Ayers,
2008; Kanoksilapatham, 2007). Information found in the four components
are summarised in Table 17.1. As the *Abstract* is short, only key steps of
each component are abstracted from the report. To help in deciding what
key steps to extract from the report for each component, suggestions are
given in Table 17.1.

There are generally two types of abstracts: descriptive and informative.
A *descriptive abstract* describes the work being abstracted and is usually
short, less than 150 words. A descriptive abstract indicates the type of
information found in the report and includes the keywords found in it. It
may include the objective and method of the research. A descriptive abstract
does not provide results or conclusions nor evaluate the research.

Table 17.1 Components of an *Abstract* and their relevant tenses

Component	Description	Tenses
1	Introduction	
	A. Research background	Simple past or present perfect
	B. Research problem	
	C. Research purpose	Simple past or present perfect
		Simple past or present
2	Methods	
	A. State data collection instruments	Simple past and past perfect
	B. Identify experimental variables	
	C. State experimental steps taken	
	D. State data analysis processes	
3	Results and Discussion[a]	
	A. Present main findings	Simple past
	B. Comment on findings as an explanation or generalisation with comparison with prior findings, or support from theories	Simple present and present perfect
4	Conclusion	
	A. State theoretical/research contributions of the study	Simple present
	B. State contributions to practice—real-world advice and/or applications	

[a] Kanoksilapatham (2013) combined discussion and conclusion parts together.

Informative abstracts, on the other hand, present the purpose, method, results, conclusions of the research and recommendations. It explains all the main arguments, important results and evidence in the report. Thus, you should aim to write an informative abstract for the final report or thesis. A sample informative abstract annotated with a content analysis is shown in Writing Guidelines 17.1 while another version of the same abstract which has been modified as a descriptive abstract is shown in Writing Guidelines 17.2.

Though the study's purpose and methods are described, the descriptive abstract does not give information about its results. Instead, it only indicates that results were reported in the paper. This is in contrast with the informative abstract version which states the principal finding and provides an explanation for it.

SOME KEY LANGUAGE FEATURES IN AN ABSTRACT

Verb tenses

Verb tense should also vary according to the rhetorical content. In stating the research background and research niche in Component 1 (Introduction), the simple past tense is used if you are describing research which took place in the field, while the present perfect tense is used to refer to research that

Writing Guidelines 17.1 Content analysis of modified informative version of Kek's (2014) abstract

Final year report informative abstract titled "Permeability of Compacted Soils" by Kek (2014)

More than two-thirds of Singapore's land area is covered with residual soils (**1A**). Residual soils are often compacted and used for engineering construction works due to its favourable engineering properties (**1A**). Compacted soils are by nature unsaturated (**1A**). Unlike saturated soils, permeability of unsaturated soils is a function of void ratio and degree of saturation (**1A**).
The objectives of this project were to investigate the effects on soil fabric on permeability and to evaluate the direct and indirect methods used to measure permeability (**1C**). In this project, both saturated and unsaturated coefficients of permeability of compacted residual soils were measured at the same relative compaction on the standard and modified Proctor compaction curves (**2A & 2B**). A flexible-wall permeameter was used to measure the saturated and unsaturated coefficients of permeability of the soil specimens directly while the minidisk infiltrometer and a permeability function (ACUPIM/W) based on the drying soil-water characteristic curve were used for the indirect methods (**2A**).
Differences in the saturated and unsaturated coefficients of permeability were observed for the specimens at the dry and wet of optimum of the compaction curves (**3A**). However, the differences in permeability could not be attributed solely to the soil fabric due to differences in the specimen void ratios after saturation (**3B**). Measured coefficients of permeability obtained from the flexible-wall permeameter and mini disk infiltrometer were consistent with values obtained from ACUPIM/W (**3B**). Suggestions for further improvement of the flexible-wall permeameter are also included in this report (**4B**).

ᵃ bracketed term at the end of each sentence indicates a step e.g., (**1B**) indicates step B of Component 1, (**2D**) indicates step D of Component 2 etc. as shown in Writing Guidelines 17.1.

Writing Guidelines 17.2 Content analysis of modified descriptive version of Kek's (2014) abstract

Modified descriptive version of final year report abstract titled "Permeability of Compacted Soils" by Kek (2014)

The objectives of this project were to investigate the effects on soil fabric on permeability and to evaluate the direct and indirect methods used to measure permeability (**1C**). In this project, both saturated and unsaturated coefficients of permeability of compacted residual soils were measured at the same relative compaction on the standard and modified Proctor compaction curves (**2A & 2B**). A flexible-wall permeameter was used to measure the saturated and unsaturated coefficients of permeability of the soil specimens directly while the minidisk infiltrometer and a permeability function (ACUPIM/W) based on the drying soil-water characteristic curve were used for the indirect methods (**2A**).
Comparative results between the saturated and unsaturated coefficients of permeability were reported for the specimens at the dry and wet of optimum of the compaction curves (**3A**). A discussion of the results ensued (**3B**). Suggestions for further improvement of the flexible-wall permeameter are also included in this report (**4B**).

ᵃ bracketed term at the end of each sentence indicates a step e.g., (**1B**) indicates step B of Component 1, (**2D**) indicates step D of Component 2 etc. as shown in Table 17.1.

has been actively conducted over recent years or research that has not been investigated up to the present study. When describing the importance or centrality of the research issue which is the focus of the study, the simple present tense and present perfect tense are most appropriate. In stating the research objective in Component 1 (*Introduction*), the simple past tense is used if you are stating your research aim as goals that have been completed. However, if you are writing it as a statement on what has yet to be revealed in the report or thesis, the simple present tense is used.

The simple past tense and past perfect tense are used in Component 2 (*Methods*). Presenting main findings in Component 3 (*Results and Discussion*) is in simple past tense but comments on the results are usually in simple present tense or present perfect tense. In Component 4 (*Conclusion*), simple present tense is used because it implies the currency of the research report's contributions to the research field it is situated in.

For more information on grammar in a report, see *Chapter 10— Grammar, punctuation and word usage guide.*

Evaluative language

Factual language should be used in stating the introduction, method and findings. However, the discussion of findings in Component 3 (*Results and Discussion*) could be expressed cautiously or confidently, depending on the empirical strength of the claims. Statements in Component 4 (*Conclusion*) are usually expressed with some caution, especially practical implications and applications as the author is aware of the tentativeness of his claims.

SOME CONSIDERATIONS IN WRITING AN ABSTRACT

A well-written abstract is very important as a reader uses it to judge if he or she should read the full report/article. Some conference organisers use the strength of the abstract to decide on the mode of presentation of the paper at the conference. An excellent abstract will be recommended for oral presentation while a good abstract will be recommended for poster presentation. Needless to say, a poor abstract will be recommended to the bin! An excellent abstract is almost invariably an informative abstract. An excellent abstract should also be concise providing sufficient information and containing no unnecessary information. The following is a checklist on writing an excellent abstract:

CHECKLIST FOR WRITING AN ABSTRACT

Introduction

- ☐ Why is the work done?
- ☐ What are the specific problems that motivate the work?
- ☐ What is the objective of the work?

Methods

- ☐ What was the methodology?
- ☐ What was investigated?
- ☐ What was done?
- ☐ How was it done?
- ☐ How was the analysis done?

Results and Discussion

- ☐ When the experiment was completed, was the hypothesis proved or disproved?
- ☐ What are the key results?
- ☐ What were the challenges in the experiment?
- ☐ Were there any corroborative results?
- ☐ What were the supporting theories?
- ☐ What were the limitations?

Conclusion

- ☐ What was the impact or implications of the work?
- ☐ What does it mean for others?
- ☐ Were the findings transferable?

PARTS OF FRONT MATTER

Front matter, otherwise known as prefatory parts of a report or thesis, typically consists of the following:

- Cover page
- Title page
- Abstract
- Acknowledgments
- Table of Contents
- List of Tables
- List of Figures
- List of Symbols

The *Cover Page* is the first page or cover of a hardbound report or thesis reflecting the title of your research project, your full official name, your university's name and, finally, the year in which your report or thesis is submitted or approved.

The *Title Page* comes after the cover page and shows the research project title, your full official name, a brief description of the nature of the research report such as 'A thesis submitted in (partial) fulfilment of the requirements for the Degree of (degree title), (your university's name) and the year of submission/approval'.

The *Abstract* follows the *Title Page*. The *Table of Contents* which follows is usually organised as two columns: the first column shows the names of the front matter, for example abstract, main parts (chapter/section names and sub-section headings) and supplementary parts (e.g. references, appendices); the second column shows the corresponding page numbers for the various prefatory parts and chapters. Note that front matter is paginated in lowercase Roman numerals (e.g. i, ii, iii, iv, v, vi) while the rest of the report is paginated in Arabic numerals (e.g. 1, 2, 3, 4, 5, 6). Note that numbering of pages starts from the first page of the Introduction to your report.

Acknowledgments comes before the *Table of Contents* and indicates your recognition of significant contributions of individuals and/or organisations towards your data collection and/or report writing. For example, you should acknowledge the contributions of your supervisor who has guided and advised you throughout the research process.

If you have tables in your report or thesis, you should include a *List of Tables*. This is oganised as three columns: the first column shows the numbering of tables in order of appearance; the second column shows the table captions or brief content descriptions; and the third column shows the page numbers which indicate where the tables appear in your report or thesis. Similarly, if you use figures in your report or thesis, you should have a *List of Figures* on the following page which is organised as three columns: the first shows the numbering of figures in order of appearance; the second column shows the figure captions or brief content descriptions; and the third column shows the page numbers which indicate where the figures appear in your report or thesis.

If you use technical or mathematical symbols, you can indicate them in a glossary of terms called a *List of Symbols*. This is usually organised as two columns: the first column shows the symbols while the second column states the explanations of the symbols.

See *Chapter 22—How to create a good layout* for more information on the layout of front matter.

CHECKLIST FOR AN ABSTRACT AND FRONT MATTER

Does your abstract include

☐ Information about your introduction, methods, results, discussion and conclusion of your study?

Does your front matter include:

☐ A cover page and title page stating your project title, your name, a brief description of the report as partial or fulfilment of a degree programme and university affiliation?
☐ An acknowledgment page?
☐ A table of contents detailing the prefatory parts, main parts and supplementary parts of your report and their page numbers?
☐ A list of tables and a list of figures if you have included tables and figures in your report?
☐ A list of symbols if you have included technical or mathematical symbols in your report?

INTERESTING FACTS

Particle physicists Berry, Brunner, and Popescu from H. W. Wills Physics Laboratory and Shukla from the Indian Institute of Technology discovered that neutrinos travel faster than the speed of light, contravening the speed limit proposed by Albert Einstein. Their finding was later shown to be erroneous, but not before the publication of their 2011 paper titled 'Can apparent superluminal neutrino speeds be explained as a quantum weak measurement?' Their abstract consists of only two words constructed as a response to their research question/title: 'Probably not.'

According to Kanoksilapatham (2013) who analysed a corpus of 60 abstracts of civil engineering research articles, research background is found in 36 out of 60 abstracts (60%) as the opening line of the abstract. This is followed by research purpose, which is found in 41 abstracts (68.33%). The methodology component, which consists of information on the data collection instruments, experimental variables and experimental steps, is the most frequent abstract component, occurring in 56 abstracts (93.33%). The results component is the next most frequent component—it is found in 55 abstracts (91.66%). The discussion component which Kanoksilapatham combined with the conclusion component is used in 40 abstracts (66.67%).

REFERENCES

Ayers, G. (2008). The evolutionary nature of genre: An investigation of the short texts accompanying research articles in the scientific journal Nature. *English for Specific Purposes* 27: 22–41.
Gillaerts, P., & de Velde, F. V. (2010). Interactional metadiscourse in research article abstracts. *Journal of English for Academic Purposes* 9: 128–139.

Kanoksilapatham, B. (2013). Generic characterization of civil engineering research article abstracts. *3L: The Southeast Asian Journal of English Language Studies* 19(3): 1–10.

Kek, H.Y. (2014). *Permeability of Compacted Soils* (Unpublished Bachelor of Engineering final year report). Nanyang Technological University, Singapore.

Chapter 18

Referencing

> ... appropriate citation and critique signals the espousal of appropriate values, etiquette, style and cultural savvy.
>
> Edward Eckel
> *Disciplinary Discourse in Doctoral Theses, 1998*

It is imperative to cite sources to back up your claims in the report by following an established citation system or style. Citation details are also scrutinised by your readers so do list all references accurately. Besides acknowledging your source materials by referencing, citation details also provide sufficient information for readers to trace the source materials and check that authors have cited their information accurately. Also, referencing helps to eliminate plagiarism, along with paraphrasing, summarising or quoting. The sources that you cite can be from books (including handbooks and reports), conference articles, online sources, patents, standards, theses, unpublished sources and periodicals.

You should also check that the sources that you are referencing are creditable. Be especially wary of online sources including online open-access periodicals. There has been an increase in low quality academic publishers providing online open-access periodicals of dubious quality. For a listing of such publishers and periodicals, see http://scholarlyoa.com/, a website maintained by an academic librarian named Jeffery Beall.

FEATURES OF TWO MAIN REFERENCING SYSTEMS

Two main systems of referencing are used in engineering documentation, namely, the author/year system and number system. The choice of the system is determined by either the institution or publication editorial board. Follow the system which the institution or publication dictates. In your report or thesis, use only one referencing system (either author/year or number) consistently. In this chapter, the APA (American Psychological Association) style is used to exemplify the author/year system while the IEEE (Institute of Electrical and Electronics Engineering) style is used to illustrate the number system. You should follow precisely the guidelines provided by the institution or publication. Pay attention to the abbreviation of the authors' names, the separators between the names, year, title, name of publication, volume number, pages and the font type.

IN-TEXT CITATIONS

In APA style, the surname of the author can serve as the subject of the sentence followed by the publication year in parentheses, or the author's surname and publication year are placed in parentheses and attached at the end of the sentence. The first type of citation is referred to as an integral citation while the second is referred to as a non-integral citation. An example of an integral citation within a sentence is 'Leong and Rahardjo (2012) …'; a non-integral in-text citation at the end of a sentence is '… (Leong and Rahardjo, 2012).' These are exemplified in Table 18.1.

In IEEE style, each reference is identified as a unique number in a square parentheses, for example, [1]. Each reference is numbered in order of appearance in the report or thesis. If you want to cite a citation more than once in the report or thesis, its unique number is used again. You can choose to state the author or lead author's surname as part of the sentence with the corresponding unique number of the source. For example, 'Leong et al. [1] measured shear and compression wave velocities of soil using bender–extender elements.' This is exemplified in Table 18.1.

The following tables point out differences in citation conventions of both systems, based on the 6th edition of the *Publication Manual of the American Psychological Association* by the American Psychological

Table 18.1 In-text citation conventions of the two referencing systems

	Author/year system	Number system
Five or less authors	Within a sentence: Leong and Rahardjo (2012) … At the end of a sentence: … (Leong and Rahardjo, 2012).	[1]
Six or more authors	Rahardjo et al. (2009) for Rahardjo, Santoso, Leong, Ng, Tam and Satyanaga (2009) where et al. is Latin for 'and others' and is used when the number of authors is six or more. Also, et al., when paired with the lead author's surname, can be used as subsequent mentions of papers by three to five authors after the first mention of all authors' surnames.	[1] or Leong et al. [1]

Association (2010) and the *2014 IEEE-SA Standard Style Manual* by the IEEE Standards Association (2014).

END-OF-TEXT CITATIONS

In APA style, all references are listed according to alphabetical order of the surnames of the authors. In a reference where there are two or more authors, only the surname of the lead author is used to list it.

In IEEE style, all references are listed according to their order of appearance, from the lowest number [1] to the highest number which represents the last reference to appear in the report.

Examples of print and non-print references in both systems are shown in Tables 18.2 through 18.6.

For an in-depth account of the APA style of referencing, refer to the *Publication Manual of the American Psychological Association* or the American Psychological Association website at http://www.apastyle.org/. Similarly, for an in-depth account of the IEEE style of referencing refer to the *2014 IEEE-SA Standards Manual* or the IEEE website at http://www.ieee.org/documents/ieeecitationref.pdf.

Table 18.2 Reference list: Author(s) of journal articles

	Author/year system	Number system
Single author	Tsakalakos, L. (2008). Nanostructures for photovoltaics. *Materials Science and Engineering R, 62,* 175–189.	L. Tsakalakos. "Nanostructures for photovoltaics". *Mater. Sci. Eng. R,* vol. 62, pp. 75–189, Nov. 2008.
Two authors	Leong, E. C., & Rahardjo, H. (2012). Two and three-dimensional slope stability reanalyses of Bukit Batok slope. *Computers and Geotechnics, 42*(1), 81–88.	E. C. Leong and H. Rahardjo. "Two and three-dimensional slope stability reanalyses of Bukit Batok slope". *Computers and Geotechnics,* vol. 42, pp. 81–88, May 2012.
Three to seven authors	Leong, E. C., Cahyadi, J., & Rahardjo, H. (2009). Measuring shear and compression wave velocities of soil using bender–extender elements. *Canadian Geotechnical Journal, 46*(7), 792–812.	E. C. Leong, J. Cahyadi, and H. Rahardjo. "Measuring shear and compression wave velocities of soil using bender–extender elements". *Canadian Geotechnical J.,* vol. 46, pp. 792–812, Jul. 2009.
In-press article	He, X., Fu, C., & Hägg, M. (in press). Membrane system design and process feasibility analysis for CO_2 capture from flue gas with a fixed-site-carrier membrane. *Chemical Engineering Journal.*	X. He, C. Fu and M. Hägg, "Membrane system design and process feasibility analysis for CO_2 capture from flue gas with a fixed-site-carrier membrane," *Chem. Eng. J.,* in press.

Table 18.3 Reference list: Books

	Author/year system	Number system
Book	Blight, G. E., & Leong, E. C. (2012). *Mechanics of Residual Soils* (2nd ed.). Leiden: CRC Press/Balkema.	G. E. Blight and E. C. Leong. *Mechanics of Residual Soils,* 2nd ed. Leiden: CRC Press/Balkema, 2012.
Edited book	Draelos, Z. D. (Ed.). (2012). *Cosmetic Dermatology: Products and Procedures.* Chichester: Wiley-Blackwell.	Z. D. Draelos, Ed., *Cosmetic Dermatology: Products and Procedures.* Chichester: Wiley-Blackwell, 2012.
Article or chapter in an edited book	Liu, M., Jiang, X. D., Kot, A. C., & Yap, P. T. (2011). Application of Polar Harmonic Transforms to Fingerprint Classification. In C. H. Chen (Ed.), *Emerging Topics in Computer Vision and Its Applications* (pp. 297–312). Washington: World Scientific.	M. Liu, X. D. Jiang, A. C. Kot and P. T. Yap. "Application of Polar Harmonic Transforms to Fingerprint Classification". In *Emerging Topics in Computer Vision and Its Applications,* C. H. Peters, Ed. Washington: World Scientific, 2011, pp. 297–312.

Table 18.4 Reference list: Other print sources

	Author/year system	Number system
Conference paper in print proceedings	Rahardjo, H., & Leong, E. C. (2006). Suction measurements. In G. A. Miller, C. E. Zapata, S. L. Houston (Eds.), *Unsaturated Soils 2006.* Paper presented at 4th International Conference on Unsaturated Soils, Phoenix, AZ (pp. 81–104). Reston: ASCE.	H. Rahardjo and E. C. Leong. "Suction Measurements," in *Proc. 4th Int. Conf. on Unsaturated Soils,* Phoenix, AZ, 2006, pp. 81–104, Apr. 2006.
Dissertation, unpublished	Aung, A. M. W. (2013). Surface wave inversion and detection of underground objects (Unpublished doctoral thesis), Nanyang Technological University, Singapore.	A. M. W. Aung. "Surface wave inversion and detection of underground objects". Ph.D. thesis, Nanyang Technological University, Singapore, 2013.
Government document	BCA (2012) *Handbook to BC1: Use of Alternative Structural Steel to BS 5950 and Eurocode 3.* [Singapore]: Building & Construction Authority.	BCA. *Handbook to BC1: Use of Alternative Structural Steel to BS 5950 and Eurocode 3.* [Singapore]: Building & Construction Authority, 2012.

(Continued)

Table 18.4 (Continued) Reference list: Other print sources

	Author/year system	*Number system*
Technical report	Tsuo, Y. S., Menna, P., Wang, T. H., & Ciszek, T. F. (1998). *New opportunities in crystalline silicone R&D* (#CP-590-25612). Retrieved from http://www.osti.gov/bridge/ servlets/purl/6593aQxa5y/ native/6593.pdf	Y. S. Tsuo, P. Menna, T. H. Wang, T. F. Ciszek, New opportunities in crystalline silicon R&D, NREL Rep, no. #CP-590-25612, 1998. [Online]. Available: http:// www.osti.gov/bridge/servlets/ purl/6593aQxa5y/native/6593. pdf. Accessed: Jan. 8, 2015.
Patent	Mei, T., & Yang, T. (2012). *U.S. Patent No. 7,898,187.* Washington, DC: U.S. Patent and Trademark Office.	T. Mei and T. Yang, "Circuit and method for average-current regulation of light emitting diodes," U.S. Patent 7 898 187 B1, 2011, Mar. 1, 2012.
Standard	International Organization for Standardization and International Electrotechnical Commission (1994). *Information technology— vocabulary—part 23: Text processing* (ISO/IEC 2382-23: 1994). Retrieved from ISO Online Browsing Platform.	*Information technology— vocabulary—part 23: Text processing,* ISO/IEC 2382-23, 1994.

Table 18.5 Reference list: Electronic sources

	Author/year system	*Number system*
Electronic book	Bass, L., Clements, P., & Kazman, R. (2012). *Software architecture in practice* (3rd ed.). Retrieved from Safari: http://proquest. safaribooksonline.com/	L. Bass, P. Clements and R. Kazman, *Software architecture in practice* (3rd edition). Addison–Wesley International, 2012. [E-book]. Available: http:// proquest.safaribooksonline. com/[Accessed: 20 Oct 2014]
Chapter/ section of a web document or online book chapter	Jones, N. A., & Gagnon, C. M. (2007). The neurophysiology of empathy. In T. F. D. Farrow & P. W. R. Woodruff (Eds.), *Empathy in mental illness.* Retrieved from: http://ebooks.cambridge.org/ ebook.jsf?bid = CBO9780511543753	N. A. Jones and C. M. Gagnon, "The neurophysiology of empathy," in *Empathy in mental illness,* T. F. D. Farrow and S. R. Campbell, Eds. 2007, pp. 217–238. Cambridge University Press. [E-book]. Available: http://ebooks. cambridge.org/ebook.jsf?bid = CBO9780511543753 [Accessed: 20 Oct 2014]
CD/CD-ROM/ DVD	Banner Engineering (2008). *Banner sensors graphical user interface version 1.2.* [CD ROM]. Minneapolis, MN: Banner Engineering.	Banner Engineering, Banner sensors graphical user interface version 1.2. [CD ROM]. Minneapolis, MN: Banner Engineering; 2008.

Table 18.6 Personal communication

	Author/year system	*Number system*
Personal communication (includes interviews, emails, letters and verbal communication)	*No personal communication is stated in your reference list. However, you state it as as an in-text citation only, for example: Within a sentence: K. K. W. Ong argued that APA style is easier to use (personal communication, October 7, 2013) At the end of a sentence: ... (K. K. W. Ong, personal communication, October 7, 2013)	*No personal communication is stated in your reference list. However, if the author is identifiable and willing to be identified, you can mention within the text, for example: In a personal interview with Dr Kenneth Keng Wee Ong, he argued that IEEE style is easier to use.

TIPS FOR SORTING REFERENCE LIST

The references are listed according to alphabetical order for the APA style of referencing and the number order of appearance in the text for the IEEE style of referencing. References are added to the reference list as you are drafting various chapters of your report or thesis. When typing the reference list in a Microsoft Office Word document, it is easier to add the reference at the end of the list so as not to interrupt your thoughts in the writing process. Such a reference list is unsorted (not in the proper order).

For the APA style of referencing instead of 'cutting' each reference and 'pasting' it at the correct location in the list, you can let Microsoft Office Word do the sorting. First, select the entire list and then select the *Sort Text tool* in the *Paragraph group* under the *Home tab* as shown in Figure 18.1. Select *Sort by paragraphs, Type: text, No header* and select *OK*. Your list of references will be sorted by alphabetical order.

For the IEEE style of referencing, use the APA style of referencing first when you are drafting the report or thesis. Sort the list of references as described in the paragraph above. When the draft is in the final form, print the hardcopy for a final check. In the hardcopy, number the references in the order of appearance. Then go to the list of references and number the references accordingly. After the report or thesis has been revised the draft, sort the list of references in Microsoft Office Word. First, select the entire list and then select the *Sort Text tool* in the *Paragraph group* under the *Home tab*. Select *Sort by paragraphs, Type: text, No header*. Select the *Options . . .* button to bring out the *Options dialog box* as shown in Figure 18.2,

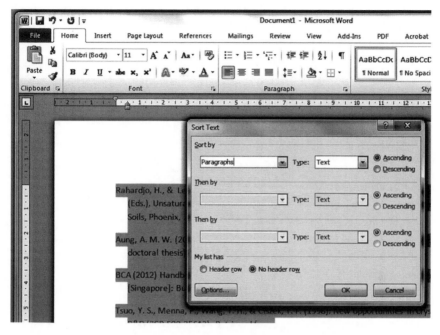

Figure 18.1 Sort Text dialog box in Microsoft Office Word

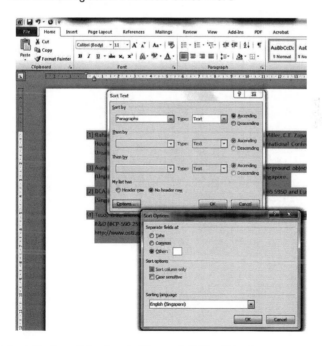

Figure 18.2 Sort Options dialog box in Microsoft Office Word

select *Separate fields at Other* and enter a space and select *OK to exit the Options . . . dialog box* and then select *OK* to sort. Your list of references will be sorted by number order.

INTERESTING FACTS

There are several referencing styles. We have only covered two, APA and IEEE, in this chapter. General information on other referencing styles is given below:

Chicago Manual of Style is the most widely used style guide by academic publishers. It is widely used in the arts and humanities. *The Chicago Manual of Style* is available online: http://www.chicagomanualofstyle.org/.

The *CSE* (Council of Science Editors) style manual uses a numbered referencing system, where the reference list is arranged alphabetically by author and numbered accordingly. It is widely used in the life sciences but is also applicable to other scientific disciplines. The manual is available online: http://www.scientificstyleandformat.org/Home.html.

Harvard is a generic term for any style that contains author-date citation in the text of the document and a list of references at the end of the document, arranged by authors' names and year of publication. There is no official Harvard style guide. The APA style is a Harvard reference style. It is used widely in academic journals.

MLA (Modern Language Association) referencing uses Harvard-style references in the text of the document, but without the year of publication. It is widely used in the fields of modern literature and linguistics. The MLA style is published as the *MLA Style Manual and Guide to Scholarly Publishing* and *MLA Handbook for Writers of Research Papers*.

Vancouver is the equivalent of Harvard for a numbered reference list. It is widely used in the health sciences. There is no official Vancouver style manual. The US National Library of Medicine's style guide is considered the most authoritative manual on the Vancouver style of referencing. The IEEE style is a Vancouver reference style.

REFERENCES

American Psychological Association. (2010). *Publication Manual of the American Psychological Association* (6th ed.). Washington, DC: American Psychological Association.

IEEE Standards Association. (2014). *2014 IEEE-SA Standards Style Manual*. New York: IEEE.

Chapter 19

Using sources and avoiding plagiarism

If I have seen further than others, it is by standing upon the shoulders of giants.

Isaac Newton (1643–1727)

USING SOURCES

In research writing, you often need to include material from published sources in your own writing. There are several reasons why you need to incorporate other people's work in your report or thesis:

- To contextualise your research by providing background information that your readers need to understand your study. This is usually done in the *Introduction* section.
- To show that you are familiar with the important research in your field.
- To establish your study as one link in a chain of research.
- To support the claims you make and thus strengthen your argument.

Types of sources

We can divide sources into primary, secondary and tertiary sources. *Primary sources* consist of original materials, created during the time of the study. *Secondary sources* consist of interpretations and analyses of primary sources. *Tertiary sources* are syntheses of primary and secondary sources. Sometimes, a document contains both primary and secondary source materials. For example, the literature review of a report or thesis contains secondary source information. However, the main body of a report or thesis which describes the experiments and findings of the study are primary sources. Generally, theses, reports, original research papers and patents are considered primary sources. Examples of secondary sources include textbooks, (also considered tertiary) review papers, databases and magazine articles. Fact-books, guidebooks, manuals and textbooks are examples of tertiary sources. When citing, you should try to cite from primary sources as much as possible.

Evaluating sources

Reliability of sources has become a very serious issue with the availability of the Internet and advent of open-access publications. Information gleaned from the Internet should always be verified for correctness and accuracy. The open-access publications' business model is based on fee collection from the authors. This has led to a rise in predatory publications which publish articles of dubious quality as long as the authors are willing to pay for their work to be published. A list of such publishers and publications can be found at http://scholarlyoa.com/publishers/.

Acknowledging sources

Wherever you have used ideas or information from other people's research in your report or thesis, you must acknowledge their contributions. This indicates respect for the intellectual property of other researchers. If you use their work without acknowledgment, you are essentially passing their work off as your own. This is called plagiarism, which is considered a very serious offence at universities and other institutions of learning. If proven, this can result in severe penalties being imposed such as a failing grade or even expulsion from an institution.

When acknowledging sources, you will need to use a proper referencing system. In the engineering field, several different referencing systems are used. An example of commonly used referencing systems is IEEE. Check with your department or supervisor the referencing style they prefer. Then it is just a matter of familiarising yourself with the referencing style you are expected to use (see *Chapter 18—Referencing* for more guidance).

AVOIDING PLAGIARISM

It is generally assumed that the words or ideas in your writing are your own unless it is indicated otherwise through appropriate citation and referencing. You need to acknowledge the source of an idea or an item of information unless it is general or field-specific common knowledge.

Field-specific common knowledge is 'common' only within a particular field or discipline. It may include facts, theories or methods that are familiar to readers within that discipline (University of Wisconsin-Madison, 2014). For instance, you may not need to give a source for your description of a testing method commonly used in engineering—but make sure that this information is so widely known that it will be familiar to your readers.

If in doubt, be cautious and cite the source. Remember also that if you use the exact words of the reference source, whether that is general or field-specific common knowledge, you must use quotation marks and acknowledge the source.

The obligation to credit the source from which you have taken ideas or information applies to any kind of source—printed, recorded, oral and online. Plagiarism of online sources (e.g. website, database, blog, podcast) is very easy to detect as institutions commonly require students to submit assignments, reports and theses through plagiarism detection software.

Three methods of integrating sources into your writing

Quoting, paraphrasing and *summarising* are the three common methods you can use to incorporate materials from published sources into your writing. Table 19.1 shows the differences among these methods.

Note: In engineering and science research writing, it is not common at all to use quotations. For this reason, this chapter will focus on paraphrasing and summarising, which are the common methods of using sources in these fields.

How to paraphrase

Paraphrasing is best limited to short chunks of text. It involves (1) changing the words and (2) changing the structure of the sentences without changing the original meaning of the text.

Table 19.1 Methods of incorporating sources

Paraphrasing	Summarising	Quoting
• Does not use exact words of source	• Does not use exact words of source	• Uses the exact words of source
• Involves writing a paragraph or section in your own words	• Involves writing only the main ideas in your own words, and omitting details	• Usually involves a brief piece of the original text
• Changes the words or phrasing in a text while retaining the original meaning	• Changes the words or phrasing in a text while retaining the original meaning	• Quotation marks used both before and after the quoted words
• Is usually the same length as the original text	• Is usually much shorter than the original text	• Same length as original text
• Source is acknowledged and referenced	• Source is acknowledged and referenced	• Source is acknowledged and referenced

Writers who specialise in writing academic English such as Gillet et al. (2009) recommend that a systematic procedure be followed in paraphrasing as shown below.

Procedure for paraphrasing

1. Read and re-read the original text until you fully understand it.
2. Identify the important ideas/main points and highlight them by underlining or using a highlighter.
3. Make a list of the highlighted ideas, paraphrasing them as much as possible.
4. Put away the original text. Imagine you are speaking about the text to someone who has not read it. Using just the main points in your notes, write your paraphrase.
5. Check your paraphrased version against the original text to make sure that the meaning in your version is the same as in the original and that all the essential information is included and expressed differently.
6. Check that your paraphrase is about the same length as the original.
7. Check that you have acknowledged the source by appropriate referencing.

In paraphrasing, the language used in the original text should be changed substantially. The procedure you can use to do this is described below.

Procedure for changing the words in the original text

1. Underline any keywords that can be changed.
2. Find other words and phrases that have similar meanings that can be used to replace the keywords in the original text. Use a thesaurus or dictionary, or look for synonyms in your word processing software. Beware, however, that some words have no true synonyms, for example *technology, energy and ductility.*
3. Circle the specialised words, that is the words that relate to the topic of the text. These will have to be included in your paraphrase, as without these words, the meaning of the paraphrase can change.
4. Change the class that a word belongs to. For instance, if it is a noun, change it to a verb and vice versa: *discussion (noun)—discuss (verb).* Or, if a word is an adjective, change it to a verb or adverb: *mechanical (adjective)—mechanise (verb); mechanically (adverb).*

Besides changing the words used in the original text, restructure the sentences. Here is how you can do it

Changing the structure of the original text

1. Break up long sentences and combine short sentences.
2. Change the word order in the sentence:

> Example:
>
> Original sentence:
>
> Scour induced by propeller or thruster wash has become one of the most important issues for the design and maintenance of navigation channels and harbour structures (Hong et al. 2013).
>
> Paraphrased sentence (word order changed):
>
> A primary consideration in designing and maintaining navigation channels and harbour structures concerns scour resulting from propeller or thruster wash (Hong et al. 2013).

Useful expressions to use when paraphrasing

The following are some expressions you can use when you are referring to someone's work when you are paraphrasing (Gillet et al., 2009):

> *The study/research by X shows/indicates/reveals that …*
> *As X has indicated/demonstrated …*
> *X has drawn attention to …*
> *X reports/argues/concludes that …*
> *A study by X shows that …*
> *Research by X suggests that …*

Summarising

Paraphrasing is usually used together with summarising when citing sources in a research paper or report. However, while paraphrasing aims to *restate* relevant information, summarising attempts to *reduce* information to a shorter length by retaining only key points and omitting details. Keep in mind, however, when you summarise for a research paper or report, you do not have to include every main point in the source or even in the section you are summarising. You should only include those points relevant to the argument you are making in your work as long as the points you omit do not change the meaning of the original text.

Ask yourself the following questions before you summarise:

1. Can readers understand the source without knowing the details? If readers need more than just the gist of what is stated in the source, it is advisable to paraphrase rather than summarise.

2. How is the information from the source relevant to my argument/the claim I am making in my report or thesis? What claim or reason does the information in the source support? How does it strengthen my argument/case/claim?

Follow the steps below to ensure that your summary is relevant and accurate (Bailey, 2011):

1. Read the original text carefully to understand what the writer is saying.
2. Highlight key points by underlining or using a highlighter.
3. Make notes of the key points and paraphrase them.
4. Write the summary from your notes, changing the structure of the original text.
5. Check the summary to ensure it is accurate and nothing important has been changed or omitted.
6. Check that you have acknowledged the source by appropriate referencing.

Original version:

There are a number of ways of expressing the total amount of water in the oceans. Seawater covers 361 million square kilometres (361 × 10^6 km²) which represents 71% of the surface of the globe. The total volume of water is enormous: 1.37 thousand million cubic kilometres (1.37 × 10^9 km³). Most of this water is contained in the three great oceans of the world: the Indian, Pacific and Atlantic Oceans (Holden, 2008).

Summary:

Most of the earth's surface (71%) is covered by seawater contributed mainly by the Indian, Pacific and Atlantic Oceans.

CHECKLIST FOR AVOIDING PLAGIARISM

☐ Have you avoided copying word-for-word material from books, the Internet or from other sources?
☐ Have you substantially rewritten the material you incorporated into your report/thesis and cited the source?
☐ Is the meaning of your paraphrase or summary the same as in the original?
☐ Have you acknowledged other people's work through appropriate referencing?
☐ Is the source of each figure cited if this is not created by you?

INTERESTING FACTS

The consequences of plagiarism are grave:

In 2008, Shamim 'Chippy' Shaik, the former head of acquisitions for the Armaments Corporation of South Africa, had his mechanical engineering PhD revoked because his thesis contained substantial amounts of plagiarised materials.

In 2011, Karl-Theodor zu Guttenberg, Germany's defence minister, quit over a plagiarised doctorate thesis in law (Causa Guttenberg). In 2013, Annette Schavan, Germany's education minister, resigned when she was stripped of her doctorate because of plagiarism (Schavanplag).

In 2013, Dongqing Li, a nanotechnology expert based in Waterloo, Canada and his graduate student Yasaman Daghighi were forced to retract an article that they published in 2010 due to plagiarism.

A more serious offence than plagiarism is fraud. A vivid account of how Jan Hendrik Schön fabricated experimental results and fooled the scientific community is described in 'Plastic Fantastic: How the Biggest Fraud in Physics Shook the Scientific World' written by Eugenie Samuel Reich.

REFERENCES

Bailey, S. (2011). Paraphrasing. In *Academic Writing: A Handbook for International Students* (3rd ed.). New York: Routledge.

Gillett, A., Hammond, A., & Martala, M. (2009). Working with other people's ideas and voices. In *Successful Academic Writing*. Harlow, Essex: Pearson Education.

Holden, J. (2008). *Physical Geography and the Environment* (2nd ed.). Harlow, Essex: Pearson Education.

The University of Wisconsin – Madison (2014). *The Writer's Handbook: Avoiding Plagiarism*. Retrieved from: http://writing.wisc.edu/Handbook/QPA_plagiarism.html.

Chapter 20

Revising and editing

Writing is largely a matter of application and hard work, of writing and rewriting endlessly until you are satisfied that you have said what you want to say as clearly as simply as possible. For me that usually means many, many revisions.

Rachel Carson
Silent Spring, 1962

Revising is the final stage of the writing process. Be sure to set aside ample time for revising and editing your report or thesis in your planning schedule. Careful revision will ensure that the document you submit is clear, concise and error-free.

The terms *revising, editing,* and *proofreading* are often used interchangeably but they should be differentiated as their focus is different as explained below [University of Toronto (n.d.), *The Writing Process*].

- *Revising* focuses on the more global aspects of writing and includes content, organisation, paragraph development and logical progression of information and ideas.
- *Editing* focuses on making changes at the sentence level. It involves looking at each sentence carefully, and making sure that it is well structured and conveys its meaning clearly and concisely.
- *Proofreading* involves checking for errors—grammatical, punctuation, spelling, typos and so on.

GENERAL ADVICE BEFORE YOU REVISE

Set aside plenty of time for rewriting

Leave plenty of time for revising your draft as this process usually takes much longer than expected. The longer your document, the more time you should set aside. If you do not give yourself adequate time to revise your report or thesis, you may find yourself having to submit an early or even your initial draft.

Avoid revising from memory

It is difficult to see the errors in our own writing. This is because we are too familiar with the text we have written and will tend to read from memory instead of actually reading what is on the page. Consequently, we fail to see the errors.

To overcome this, create distance between you and your document by waiting for a few days before you start revising. It then becomes easier to see errors of organisation and style after a cooling off period.

Revise on hardcopy

Studies have found that people generally miss errors in their documents when they are reading from the computer screen. As we cannot see the entire document on the screen, it is difficult to notice structural problems. Revising on a hardcopy of your report or thesis will not only give you a better sense of the structure of your document but will also enable you to catch more errors. For revisions on hardcopy, writers use editing symbols to indicate the changes that are to be made. Your supervisor may also use editing symbols when he reads your draft report or thesis. Common editing symbols are given in the Appendix.

Adopt the reader's perspective

In revising, read your writing from the reader's point of view. Think as a critical reader. Read every sentence to see if it conveys the message you intend and how your readers may respond to the claim made. That way, you can anticipate questions from the reader and rewrite appropriately.

HOW TO REVISE

Silyn-Roberts (2013) recommends a 'multi-pass' process of revising in which different aspects are revised at each pass going from overall organisation to details.

To revise efficiently, revise in stages, focusing on only certain aspects in each stage. This may sound time-consuming, but it has been shown to take less time as well as being more effective than trying to revise everything at the same time.

Advice on how to edit efficiently recommends that you revise 'top down' or from whole to part. In this approach, you begin revision by looking at overall organisation, then sections, paragraphs and sentences, and finally grammar, spelling and punctuation. Table 20.1 shows a possible scheme that goes from overall organisation to details.

Table 20.1 Stages of revision

Stages	Organisation
Stage 1: Revising	• Content • Overall organisation • Paragraph structure
Stage 2: Editing	• Style • Sentence length and structure
Stage 3: Proofreading	• Grammar • Punctuation • Spelling, typos
Stage 4: Formatting	
Stage 5: Document integrity	

Stage 1: Revising—content, overall organisation and paragraphs

Checking content

As revising content may significantly alter parts of a document, it is advisable to review and revise content thoroughly before beginning to edit for style, usage, grammar, punctuation and spelling.

Questions about content

Read through your draft slowly, stopping at the end of each section, and ask yourself the following questions:

1. *Is the information accurate?* Correct any inaccurate quantitative data or misleading graphics.
2. *Is the information complete?* Fill in gaps in information in your draft and add additional information if necessary.
3. *Is the information comprehensible and logical?* Rewrite any information, explanation or description that is unclear, ambiguous or confusing.
4. *Is any irrelevant information included in the document?* Delete any information that is unnecessary, tangential or unimportant to your readers. (If the information is necessary only for some readers, include it in one or more appendices.)
5. Are all technical terms that need to be defined clearly defined? Are technical terms used correctly and consistently?

Checking overall organisation

Read your document critically in an overall way. Look only for errors of organisation.

1. Does your draft match your outline?
   If not, why not? Does your outline need to be revised, or would your original structure be better?
2. Is the structure logical?
   To check if your document is logically structured, print it out. Next to each paragraph, write a phrase or sentence that captures the key point of the paragraph in the margin. Then read only the phrases or sentences that you wrote; do they flow logically? If not, this indicates a flaw in the structural order of the document. To fix that, determine if the sections or paragraphs are in the most effective order. Should this paragraph come before that one, that section before this one?
3. Is the information coherent?
   To check if your document is coherent, look at the points where you move between ideas in your writing. These points are referred to as 'transitions' and occur between sections and paragraphs as well as within sentences and within paragraphs. Identify and evaluate the strategies you use for transitioning. Transitional strategies play an important role in ensuring coherence or 'flow' in your document. Common transitional strategies include:
   Logical transition: The last idea of the previous section or paragraph is the first idea of the next section or paragraph.
   Phrasal transition: Use explicit wording to create a shift in writing (e.g. *first ..., second..., next...*); develop a relationship between ideas in sentences/paragraphs (e.g. *because, consequently, in contrast* etc.).
   Structural transition: Use similar/parallel sentence structure to create a relationship between sentences, for example writing sentences in the passive voice to describe procedures.
   Verbal transition: Use the same key words to establish a relationship between sections or paragraphs.
   Transitional strategies create 'flow' or coherence in your document by clarifying the relationship of ideas in a piece of writing. Check that you have used transitional strategies correctly as using inappropriate ones will make your writing incoherent and confuse your readers.

Checking paragraphs

Check each paragraph for the following:

- The paragraph has a clear and accurate topic sentence near the beginning presenting the main point of the paragraph.
- Each sentence in the paragraph supports the main point.
- Sentences in the paragraph are logically organised using an appropriate organisational principle (e.g. chronology, cause and effect,

comparison and contrast, pros and cons etc.) Assess whether the organisational structure is efficient or if there is a better structure.
- The paragraph's role in its section and in the document as a whole is clear.

Avoid very long paragraphs (more than half a page) or very short ones (fewer than five lines). Reserve the use of two- or three-sentence paragraphs for introductions and conclusions of sections.

Stage 2: Editing

Editing and proofreading both focus on the sentence level. Editing involves critically examining and questioning sentences whereas proofreading only involves checking sentences for errors. When editing:

- *Check sentences to see if they are too long (more than 20 words) or too short.* Break up overly long sentences. These are usually difficult to read as well as make it difficult to see the relationship of ideas in the sentence. Very short sentences can be combined so the connection between ideas can be better seen.
- *Read each sentence carefully and check if it supports the topic sentence.* Does it need to be rewritten more effectively for accomplishing that goal?
- *Check links between sentences.* Analyse the sentences that come before and after the sentence you are focusing on. Are the connections between these sentences clear, or do you need to insert transitions between them?
- *Check sentences to see if the subject and the verb are close together* (not interrupted by more than a word or two). Every sentence has a subject (actor) and a verb (action). Once the subject is stated, the reader expects to be quickly informed about the action that the subject performs. It is difficult for the reader to process information in the sentence if the subject and verb are interrupted by a long string of words in between.
- *Check sentence structure to see if they are too elaborate.* Simplify unnecessarily complicated phrasing of ideas as in the following sentence. **Example of unnecessarily complicated phrasing:**
  It is evident that this thesis provides a methodological foundation from which engineers may undertake the betterment of the circuit board manufacturing process.
  Revised version:
  This thesis provides a method for improving the circuit board manufacturing process.
- *Eliminate long introductory phrases in sentences.* Good technical writing is concise and precise. Sentences should get to their point

quickly. In the example given earlier, the phrase 'It is evident that' serves no useful purpose and should be dropped. Other examples of unnecessary introductory phrases are 'In view of the fact that', 'It should be noted that', 'Needless to mention' and so on (see *Chapter 11—Do's and don'ts of technical writing* for other examples of meaningless introductory phrases).

Stage 3: Proofreading

When proofreading check sentences for grammar, punctuation and spelling errors.

Sentences

- Is the right tense used?
- Is there subject–verb agreement?
- Is it a 'real' sentence or is it a fragment? [A real or complete sentence has a subject, a complete verb and conveys a complete thought (see *Chapter 10—Grammar, punctuation and word usage guide* for common errors in sentences)].

Words and spelling

- Is the right word (affect/effect, led/lead, lose/loose etc.) used?
- Are contractions (isn't, hasn't, can't etc.) used?
- Are colloquialisms or informal words used?
- Are words correctly spelled?

Punctuation

- Are commas, full stops, colons and semicolons used correctly and effectively?
- Are apostrophes used correctly?

As proofreading is mainly mechanical, you can use the grammar checker or spell checker on your word processor. However, even after spell checking, do not assume that your document is error-free. Proofread it yet again. A spell checker will overlook words you have used unintentionally, for example *an* instead of *on*, *it* instead of *is*, *their* instead of *there* and so on.

Stage 4: Formatting

Check that you are following your department's guidelines for fonts, paragraph formatting, justification, spacing and margins.

Stage 5: Document integrity

The term 'document integrity' is used by Silyn-Roberts (2013, p. 200) to refer to matching of the following:

- Section numbers to *Table of Contents* page
- Page numbers in text and *Table of Contents* page
- Text references and figure numbers

In this last stage, check if

- The section headings and sub-headings are correctly numbered and consistent.
- Reference is made to the figures in the text.
- The page numbers in the *Table of Contents* correspond to page numbers in the text and the wording of the headings match those in the text.
- Illustrations are numbered, title, labelled appropriately, and correctly referred to in the text.

Check references

- For each in-text citation, check if there is a corresponding reference in the *List of References* section and vice versa.
- Check if the date of the in-text citation matches the date in the full reference in the *List of References* section.
- Check that all the references in the *List of References* section are formatted consistently and all the necessary details included.

INTERESTING FACTS

Who knows more about the art of revision than great writers? Here some famous writers share their thoughts about revising (extracted from Temple, 2015):

'I have rewritten—often several times—every word I have ever published. My pencils outlast their erasers.'
— Vladimir Nabokov, *Speak, Memory,* 1966

Interviewer: How much rewriting do you do?
Hemingway: It depends. I rewrote the ending of *Farewell to Arms,* the last page of it, 39 times before I was satisfied.

Interviewer: Was there some technical problem there? What was it that had stumped you?

Hemingway: Getting the words right.

— Ernest Hemingway, *The Paris Review Interview*, 1956

'Anyone and everyone taking a writing class knows that the secret of good writing is to cut it back, pare it down, winnow, chop, hack, prune, and trim, remove every superfluous word, compress, compress, compress...'

— Nick Hornby, *The Polysyllabic Spree*

REFERENCES

Silyn-Roberts, H. (2013). Revising. In *Writing for Science and Engineering: Papers, Presentations and Reports* (2nd ed.). Waltham, MA: Elsevier.

Temple, E. (2015). *20 great writers on the art of revision.* Flavorwire. Retrieved from: http://flavorwire.com/361311/20-great-writers-on-the-art-of-revision.

University of Toronto. (n.d.). *The writing process: Revising, editing and proofreading.* Retrieved from: http://www.engineering.utoronto.ca/Directory/students/ecp/handbook/process/editing.htm.

Part III

Presentation

Chapter 21

How to create figures

Use a picture. It's worth a thousand words.

Arthur Brisbane (1864–1936)

Figures are common in technical writing. The purpose of figures is to communicate information to the reader which writing alone cannot achieve. Figures are thus an extension of the text. However, they should be understood on their own without extensive explanation in the text. This chapter presents an introduction to the various types of figures, a guide on how to create them and tips on producing effective figures.

PURPOSE OF FIGURES

We use figures to communicate what cannot be conveyed by text alone. Each figure should deliver a key message. Figures should never be used merely to dress up or fill up a paper (this practice is called 'padding'). If a figure is superfluous, the reader will be able to sense it immediately. When reviewing journal papers, a reviewer must consider two key questions concerning figures in a paper:

1. Are all the figures necessary? (In other words, are there figures that can be deleted?)
2. Are the qualities of the figures acceptable for publication?

This chapter will address both questions.

TYPES OF FIGURES

We can classify figures into three types:

- Photographs and images
- Schematic drawings
- Graphs and charts

PHOTOGRAPHS AND IMAGES

Copyright and permission

Photographs and images are sometimes needed to make or clarify a point. Unless the photographs and pictures are produced by you, you will need to get permission for their use. Using copyrighted photographs and images without permission is considered plagiarism. You should assume that all images found on the Internet are copyrighted even though no explicit copyright statement is attached to an image. Although it is quite a common practice in academia to use a figure from a published paper and attribute it to the source based on 'fair use', the practice for images that are found on the Internet is not as clear. Copyright laws are constantly evolving. If you are in doubt about the copyrights of materials you use, it is better to err on the side of caution and seek permission for their use. In some universities, it has become a requirement for the students to seek permission for the use of all referenced images in their reports and theses.

You can find standard letters on the Internet for seeking permission for the use of copyrighted materials. All publishers will have such a standard letter. Your university may also have such a letter. The standard letter for seeking permission on the use of copyrighted materials contains the following:

1. From you (your particulars: name and address)
2. To them (their particulars: name and address)
3. To use in your publication (state type, title and publisher of your publication)
4. From the publication (state details of the publication you will be citing; this includes a declaration that you will list it in your reference and the picture will also include the 'Reproduced by permission of [copyright holder]' in the figure caption)
5. Signature and date

As you can see, seeking permission for use of copyrighted materials is quite a cumbersome and time-consuming process. So before you decide to use copyrighted materials, consider the following questions:

1. Is the use of the copyrighted material absolutely necessary in your report or thesis?
2. Are you using the copyrighted figure as it is, or are you using it as part of your figure? If you are modifying the figure extensively, you may be better off creating a completely new figure.
3. Is the copyrighted material largely generic? If it is, you can create your own figure without using the copyrighted material.

If you need to seek permission to use copyrighted materials, you should start the process as early as possible so that the writing of your report or thesis is not delayed.

Quality of photographs and images

The quality of photographs and images should be as high as possible. Although guidelines on image quality for students' reports and theses are not as high as those required by publishers, it is still a good idea for you to produce high quality images in case you use them later for journal paper publications.

Image quality is measured in terms of image type (format) as well as resolution. For photographs, the two common image types accepted by publishers are TIFF (Tagged Image File Format) and JPEG (Joint Photographic Experts Group). When you are taking photographs using a digital camera, save the image in one of these two formats. When you have a choice to save the photograph as a TIFF or JPEG image, choose the TIFF image as it contains the more complete image information.

As mentioned earlier, image quality is also measured in terms of resolution. The minimum resolution for image quality accepted by publishers is 300 dpi (dots per inch). However, when you take a photograph with a digital camera, you are given the image size in terms of pixels, for example, 1024 × 768, 1920 × 1080, 2560 × 1920 pixels and so forth. The three properties of an image (printed size, dpi and pixels) are related. To determine the maximum printed size of the photograph in inches at 300 dpi, divide the number of pixels by 300, that is, a 1920 × 1080 pixel image will give a maximum printed size of 6.4 inches × 3.6 inches at 300 dpi (1 inch = 25.4 mm).

Suggested image sizes for different numbers of photographs to be presented on a letter size (8.5 inches × 11 inches) or A4 size (8.27 inches × 11.69 inches) page at 300 dpi are given in Table 21.1.

If you have a digital camera that allows you to take a photograph at different image sizes, select the largest image size to take the photograph. The general expectation of photographs is that they should be relevant to the report or thesis content. Therefore, a digital compact camera with several

Table 21.1 Suggested image size for number of photographs on one letter- or A4-size page

No. of images on a letter- or A4-size page	Image size, pixels
1	2448 × 3264 (8 Million)
2	2560 × 1920 (5 Million)
3	2048 × 1536 (3 Million)
1 × 2	1920 × 1080 (2 Million)
2 × 2	1920 × 1080 (2 Million)
3 × 2	768 × 1024 (1 Million)

scene modes is probably all you need for taking photographs. Adhere to the following rules when taking photographs:

1. Position your subject (sample, equipment, set-up etc.) such that it is in the centre of your photograph.
2. Avoid unnecessary clutter in your photograph. Use a back screen to cover the background clutter, if necessary. For example, if you are taking a photograph of a test sample make sure that your photograph only shows the test sample and nothing else.
3. Use a tripod as far as possible. This will forestall fuzzy photographs due to unstable handling of the camera and it will also allow you to frame the subject properly.
4. Orientate your camera so that the vertical or horizontal elements of the subject in the photograph appear vertical or horizontal and not slanted.
5. Make sure your camera is in focus before taking the shot.
6. Ensure your photograph is not underexposed (too little light) or over-exposed (too much light). Normally this can be avoided using one of the automatic scene modes on a digital compact camera.
7. Ensure that you take the photograph at the highest possible image size allowed on your camera.
8. Take the same photograph several times with different scene modes and different views. You can decide later which version to use later. Remember that there are no additional costs for taking more photographs than you need with digital photography.

Software that allows you to manipulate a digital image is widely available. Ethical rules of research require that images should not be manipulated to deceive the viewer. If you follow the eight simple rules above, you should not need to manipulate a photograph. The only aspects you are allowed to adjust are brightness, contrast or colour balance as long as these adjustments do not obscure or eliminate any information present in the original photograph.

For images, the format can be in PDF (Portable Document Format), EPS (encapsulated postscript), TIFF or BITMAP (map of bits). There are other sources from which you can obtain an image. Sources include PDF files, screenshots or scanning from a hardcopy (article or page from a book). If you are copying the image from a PDF file, consider using a screen capture (*Print Screen*) of the image magnified (zoomed) to the largest possible extent on your monitor. You will find that the quality of the image obtained in this way is much better than that obtained using the *Snapshot Tool* in the Adobe Reader software. The screen capture can be directly pasted onto your report or thesis in Microsoft Word. When you click on the pasted screen capture, the *Picture Tools* tab will appear and clicking the *Format* command allows you to select the *Cropping Tool* for cropping the part of the screen capture that you do not want to be in your image. If you have acquired the image by scanning, it may be better if you first enlarge the image using a photocopier before scanning the enlarged image. When scanning, ensure that the image is positioned squarely on the scanner bed; that is, check that the scanned image does not become a distorted image of the original. You should also set the scanner settings to the highest possible resolution and save the image as a TIFF file. The quality of the image is dependent on the original image size. The minimum resolution of images demanded by publishers is 300 dpi which, as explained previously, is dependent on the image size that you want to present in your report or thesis. As a rule of thumb, the size of the image in your report or thesis should not be larger than the original size of the image. If it is, the quality of the image in your report or thesis will be poor.

SCHEMATIC DRAWINGS

Schematic drawings refer to circuit drawings, process diagrams, flowcharts for a computer program or test set-up. There are dedicated software that allow you to draw specific types of schematic drawings. If you are not making a number of schematic drawings for your report or thesis, you should just consider using the *Drawing Tools* in Microsoft Office Word. Examples of some schematic drawings using the *Drawing Tools* in Microsoft Office Word are shown in Figure 21.1.

Using the drawing tools in Microsoft Office Word

Microsoft Office Word provides a *Drawing Tools* for making drawings. To use the *Drawing Tools*, select the *Insert* tab and you will see an *Illustrations* group with *Picture, Clip Art, Shapes, SmartArt* and *Chart* commands. We shall focus on the *Shapes* command as illustrated in Figure 21.2.

The *Shapes* command offers several groups of shapes: *Lines, Basic Shapes, Block Arrows, Flowchart, Callouts, Stars and Banners* and a *New Canvas* command. These groups of shapes are used to create the schematic drawings shown in Figure 21.1. With a bit of creativity and imagination, the drawing tools can make very sophisticated schematic drawings which will satisfy the requirements of any publisher. The following are tips on how to make the best use of the drawing tools in Microsoft Office Word to make schematic drawings:

1. Before making a schematic drawing in Microsoft Office Word, first sketch the schematic drawing by hand on paper. You may have sketched the schematic drawing in the Research Log (see *Chapter 8— Keeping research records* on the advantages of recording all your research activities in a research log).
2. Begin by creating a new file in Microsoft Office Word. It is a good idea to store each schematic drawing in a new Word file with a meaningful filename so that you can find it easily. Let us assume that you

want to make a schematic drawing of the Wheatstone bridge circuit. Name this file 'Wheatstone bridge.doc' or 'Wheatstone bridge.docx', depending on the version of Microsoft Office Word you are using.

3. Select the *Insert* tab and choose the *Shapes* command. There are two ways you can draw on the new document. You can draw directly on the new document or you can first create a drawing canvas. If you

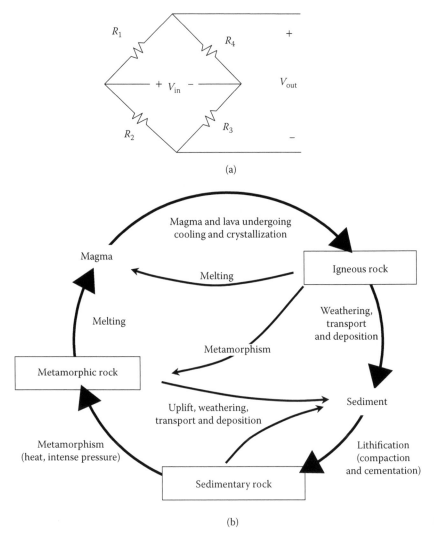

(a)

(b)

Figure 21.1 Examples of schematic drawings using Drawing Tools in Microsoft Office Word. (a) A circuit drawing of a Wheatstone bridge (circuit diagram). (b) A rock-cycle (process diagram). (c) A bubble sort algorithm to sort an array A[i ... n] (flowchart). (d) An ultrasonic test for detection of flaw (test set-up).

(Continued)

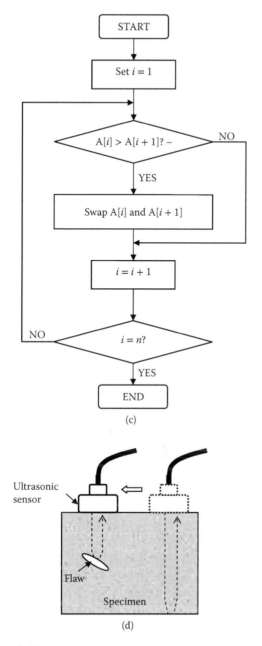

(c)

(d)

Figure 21.1 (Continued) Examples of schematic drawings using Drawing Tools in Microsoft Office Word. (a) A circuit drawing of a Wheatstone bridge (circuit diagram). (b) A rock-cycle (process diagram). (c) A bubble sort algorithm to sort an array A[i ... n] (flowchart). (d) An ultrasonic test for detection of flaw (test set-up).

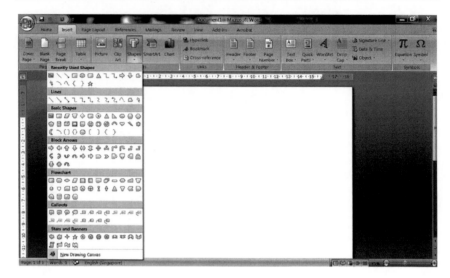

Figure 21.2 Drawing tools in Microsoft Office Word.

draw directly on the new document, the shapes will appear on the same layer as the text in Microsoft Office Word. However, the shapes may move when you edit the text around the shapes. Our preference is to first insert a new drawing canvas and draw on it. You can imagine the drawing canvas as a separate layer that is 'floating' on the text layer of Microsoft Office Word document. The drawing canvas allows you to move, copy or resize everything on the drawing canvas. If you choose not to use a drawing canvas, it is recommended that you group all the drawing objects after you have finished drawing to keep the shapes in their respective positions. Group all the drawing objects by selecting all of them, right click and select the *group* command. This will ensure that all the objects are treated as one composite object and you can then move it around the Microsoft Office Word document or copy and paste it onto another Microsoft Office Word document.

4. Treat the shapes in the drawing tools as building blocks of your schematic drawing. Select the shape that best represents the part you are drawing by clicking the left mouse button and release. Click the left mouse button on the drawing canvas and while still depressing the left mouse button, drag on the drawing canvas to draw the shape. Your schematic drawing will slowly take form as you add shapes and position the shapes on the drawing canvas.

5. To make a square with the rectangle shape or circle using the oval shape, depress the *Ctrl* and *Shift* key at the same time as you drag the mouse—this will force the shape to be either square or circle.

6. Change the size of a shape by clicking on the shape and then dragging one corner of the shape to the size you want it to be. For more

accurate sizing, you can click on the object to bring up the *Drawing Tools* tab and then select the *Format command*. The *Format command* will show a dropdown list and there will be a *Size group* which allows you to adjust the width and height of the object. You can also click on the tab on the right of the *Size group* to bring up the *Format Autoshape* dialog box as shown in Figure 21.3. Besides allowing you to change the width and height of the object, the dialog box has options to allow you to rotate the object, and adjust the size of the object proportionally with and without maintaining aspect ratios. If you want to maintain the aspect ratio of the object, check the *Lock Aspect Ratio* checkbox under *Scale* in the dialog box.

7. Align several shapes using the *Align command* in the *Arrange group*. You can select all the shapes that you want to arrange in a row or column and then select the *Align command* in the *Arrange group*. The *Align command* allows you to align all the shapes to follow a top margin, to follow a left or right margin or to centre them. Within the *Arrange group*, you can also select the *Distribute Vertically* or *Distribute Horizontally* command to make equal spacing between all the selected shapes. You can also rotate the selected objects together.

8. Exercise a sense of proportion when making a schematic drawing even though you are not expected to draw it to scale. For example, the Wheatstone bridge in Figure 21.1a looks odd when one of the resistances, R3, is drawn oversized compared with the rest of the resistances in the circuit as shown in Figure 21.4.

9. Note that the details in a schematic drawing can vary from a very simple representation to a detailed representation of the actual item

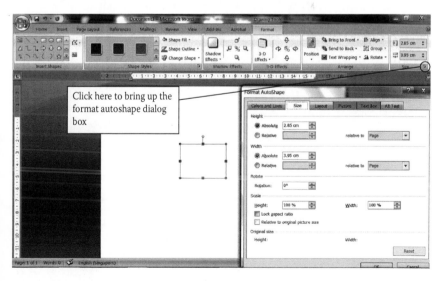

Figure 21.3 Format Autoshape dialog box in Drawing Tools of Microsoft Office Word.

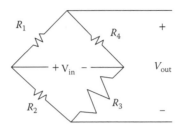

Figure 21.4 Importance of proportion in schematic drawing.

(a)   (b)

Figure 21.5 Simple and detailed representations of a function generator. (a) Simple. (b) Detailed.

that you want to show. For example, simple as well as detailed representations of a function generator are shown in Figure 21.5. When making a schematic drawing, decide on the level of detail necessary to convey your message.

10. Pay attention to the schematic drawings you see in books, papers and on the Internet. Educate your eyes on what will make a good schematic drawing. You will also need a certain amount of practice to draw good schematic drawings.

GRAPHS AND CHARTS

Graphs are a form of chart. Most of us have learned how to plot an $x–y$ graph on a piece of graph paper in elementary school as well as learned how to explain the relationship between the x and y variables using the $x–y$ graph. In Microsoft Office Excel, the $x–y$ graph is known as the $x–y$ (scatter) chart.

Microsoft Office Excel offers a variety of charts: column, line, pie, bar, area, $x–y$ (scatter), stock, surface, doughnut, bubble and radar (Figure 21.6). The most commonly used charts in engineering are the line, $x–y$ (scatter), column, bar and pie. We will focus on these five types of charts here. To illustrate these five chart types, the Changi Airport Weather Station's data from Metrological Services of Singapore for May 2012 will be used. The data are shown in Table 21.2. A brief description of the different types of charts is given in Table 21.3.

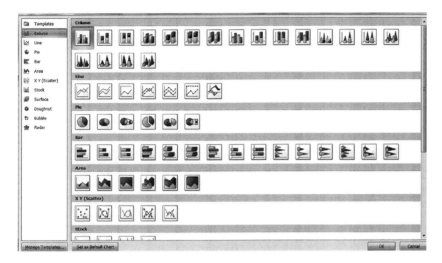

Figure 21.6 Chart types available in Microsoft Office Excel.

Table 21.2 Data from Changi Airport weather station

	Temperature (°C)			Relative humidity (%)							Surface wind		
Date	Max.	Min.	Mean	Max.	Min.	Mean	Dewpoint (°C)	Mean amount of cloud%	Sunshine duration (h)	Total solar radiation (mwh/ cm²)	Rainfall (mm)	Prevailing wind direction	Mean speed (m/s)	No. of calm hours
1	31.6	23.9	27.8	99	69	86.7	25.2	88.75	4.6	326.68	2.8	S	1.9	0
2	30.8	25.2	27.6	100	69	86.3	25	86.25	4.95	380.56	0.4	WNW	1.7	0
3	32.5	26.1	28.9	93	60	80	25	87.5	3.65	390.85	0	W	1.5	0
4	30.9	24	28	100	78	88.6	25.9	87.5	2.2	210.56	79.2	S	1.8	0
5	31.4	22	26.9	100	62	83.3	23.6	92.5	2.05	395.84	0	SW	1.6	0
6	32.3	22.1	26.8	100	63	87.6	24.4	91.25	4.6	308.05	34	VAR	1.6	0
7	28.7	22.9	26	99	70	87.9	23.8	87.5	0.2	164.72	1.2	W	1.4	0
8	33.7	25.1	28.7	98	52	80	24.6	87.5	10.3	500.56	18	VAR	2	0
9	32.6	22.6	27.4	100	57	83.7	24.2	87.5	8.25	558.89	0.4	NNW	1.9	0
10	33.1	25.2	28.2	100	52	84.5	25.1	85	8.35	611.38	0	VAR	2.2	0
11	32.1	25.3	28	100	61	86.3	25.4	87.5	6.2	395.84	10.2	WNW	1.4	0
12	33	26	29.1	100	56	79.4	24.9	86.25	9.2	472.49	18.6	W	1.6	0
13	31.6	23.1	27.6	100	65	81.9	24	93.75	0.9	335.84	0	W	2.1	0
14	33.7	26.7	29.8	93	55	74.5	24.5	87.5	10.15	584.18	0	W	1.7	0
15	31.5	25.6	28.6	95	73	84.5	25.7	87.5	2.85	285.29	22	SSW	1.1	0
16	31.9	22.9	28.2	98	63	82.6	24.9	88.75	7.05	562.51	23	SSW	2	0
17	30	23.1	27.2	100	72	86.1	24.6	93.75	0	236.4	1.4	SSW	1.6	0
18	32.1	26	29	97	64	81	25.3	87.5	7.45	529.7	0	S	2.3	0
19	32.6	26.8	29.4	93	63	79.3	25.4	86.25	9.3	489.17	0	SSE	2.6	0
20	31.3	24.3	28.5	96	63	84.4	25.5	86.25	8.35	497.78	6	ENE	2	0
21	32	26.2	28.9	99	61	83.1	25.5	87.5	7.25	528.6	0	SSW	1.7	0
22	32.1	26.1	29.3	94	59	79.9	25.3	87.5	10.25	614.44	27.2	SSW	2.2	0

(Continued)

Table 21.2 (Continued) Data from Changi Airport weather station

	Temperature (°C)			Relative humidity (%)							Surface wind		
Date	Max.	Min.	Mean	Max.	Min.	Mean	Dewpoint (°C)	Mean amount of cloud%	Sunshine duration (h)	Total solar radiation (mwh/ cm²)	Rainfall (mm)	Prevailing wind direction	Mean speed (m/s)	No. of calm hours
23	30.3	23.5	27.5	100	66	83.2	24.3	87.5	3.95	484.99	0.2	SSW	2	0
24	31.7	26.6	29	89	62	77.3	24.5	87.5	7.8	551.39	0.8	SSW	2.4	0
25	30.6	25.1	28.4	91	70	79.4	24.5	87.5	2.15	412.77	0	S	2.9	0
26	32	25.8	28.4	95	65	83.1	25.2	87.5	4.15	357.5	12	SE	3	0
27	29.4	22.8	27	99	75	86.6	24.5	91.25	1	288.61	0	SSW	1.9	0
28	31.6	25.9	28.7	98	65	81.3	25.1	86.25	8.35	481.66	0	SE	2.5	0
29	31.3	26.7	28.3	95	70	84.9	25.5	83.75	5.15	417.22	0	E	2.5	0
30	32.3	26	28.5	99	55	82.5	25	87.5	7.45	566.11	0.8	VAR	2	0
31	32.2	24.8	27.7	99	59	85.2	24.7	87.5	5.45	415	33.8	NW	1.5	0

Table 21.3 Description of chart types

Chart type	Example
The *line chart* is suitable for plotting a series of data acquired over time, for example maximum and minimum daily temperatures where the temperature is on the *y*-axis and the day is on the *x*-axis.	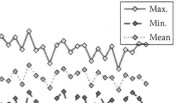
The *x–y (scatter) chart* is the most widely used chart in engineering. If you look at an engineering paper, there is a very high probability that you will see at least one *x–y* (scatter) chart. The chart expresses the relationship between two variables, *x* and *y*. The main difference between a line and an *x–y* (scatter) chart is that the *y*-data need not be spaced at regular intervals.	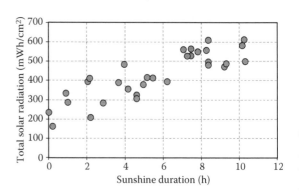

(Continued)

Table 21.3 (Continued) Description of chart types

Chart type	Example

A *column chart* compares variables such as those in the chart on the right which compares the daily rainfall for May 2012. At first glance, a column chart looks like a histogram (chart on the bottom right). Although there are differences between the two, both are represented with vertical bars. A histogram shows a distribution with the data grouped into bins or intervals as the *x*-axis whereas a column chart shows vertical bars on a category *x*-axis. The area of the bar in a histogram represents the percentage whereas the bars in a column chart are of a uniform size. In a histogram, there is no gap between the vertical bars. The bars in the histogram are ordered and cannot be rearranged unlike bars in a column chart. The histogram on the right was plotted in Microsoft Excel and does not comply fully with the rules for a histogram the area of the bar does not represent a percentage and the intervals on the *x*-axis are equal and are not labelled clearly.

A *bar chart* is similar to a column chart except that instead of vertical bars, it displays the variables using horizontal bars instead.

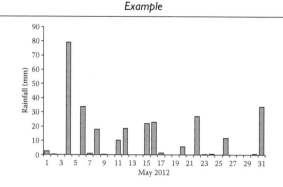

Column chart

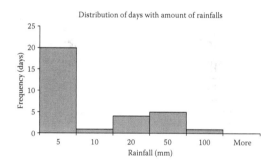

Histogram

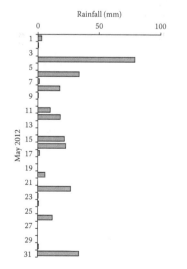

Bar chart

(*Continued*)

Table 21.3 (Continued) Description of chart types

Chart type	Example
A *pie chart* is used to show data as part of a whole. The example on the right shows the proportion of sunshine hours on 1 May 2012.	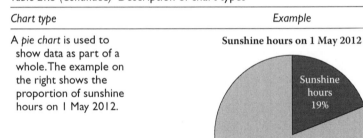 Sunshine hours on 1 May 2012 Sunshine hours 19% 81%

Creating charts in Microsoft Office Excel

To create the charts in Table 21.3, the data shown in Table 21.2 are keyed into Microsoft Excel as shown in Figure 21.7.

To create a line chart:

1. Select the data maximum, minimum and mean temperatures from cells B3 to D34. Row 34 is the last row of the data.
2. On the *Insert tab*, in the *Charts group*, click *Line*. Select the chart with markers as indicated in Figure 21.8. You will obtain a line chart that is not the same as the line chart you see in Table 21.3.

Figure 21.7 Input of Table 21.3 into Microsoft Office Excel.

3. To get the same line chart in Table 21.3, you will need to do additional formatting by changing the attributes of the line chart. This will be explained in the section on *Changing attributes of a chart*.

To create the *x–y* (scatter) plot:

1. Select the 'Sunshine duration' and 'Total solar radiation' from cells J3 to K34.
2. On the *Insert tab*, in the *Charts group*, click *Scatter*. Select the chart with markers as indicated in Figure 21.9. You will obtain an

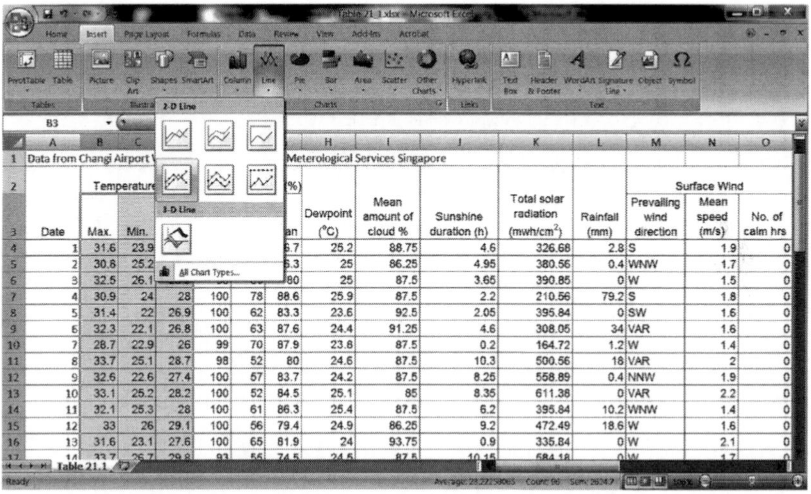

Figure 21.8 Plotting a line chart in Microsoft Office Excel.

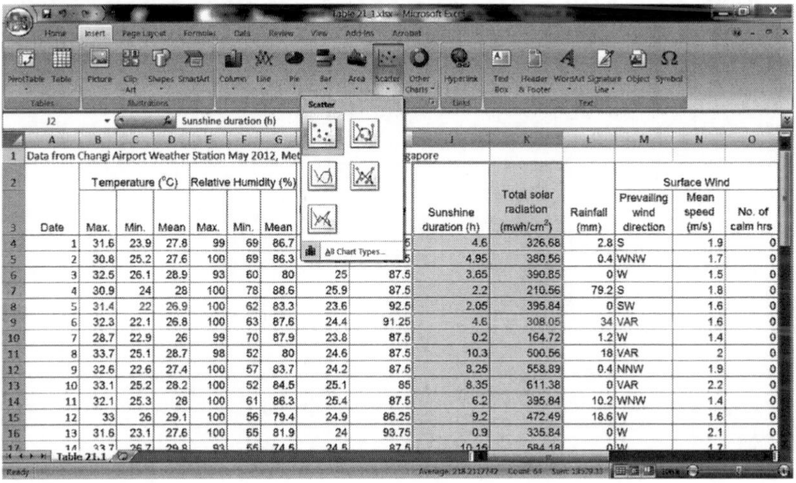

Figure 21.9 Plotting an *x–y* (scatter) chart in Microsoft Office Excel.

x–y (scatter) chart which is different from the x–y (scatter) chart in Table 21.3; this is the chart after additional formatting (see section *Changing attributes of a chart*).

To create a column chart:

3. Select the rainfall column from cells L3 to L34.
4. On the *Insert tab*, in the *Charts group*, click *Column*. Select the 2D column chart with individual vertical bars as indicated in Figure 21.10. You will obtain a column chart of rainfalls for each day of the month of May. You will need to do additional formatting by changing the attributes of the column chart to make the column chart look like that shown in Table 21.3. This will be explained in the section on changing attributes of a chart.

To create a histogram in Microsoft Excel:

1. Check first if the *Data Analysis ToolPak* is installed in your Microsoft Excel. On the *Data tab*, you should see an *Analysis group* with a *Data Analysis command* as shown in Figure 21.11.
2. If the *Analysis group* is missing or the *Data Analysis command* is missing, install the *Data Analysis ToolPak*. This can be done by (1) clicking the Microsoft Office button  , and then clicking the *Excel Options*; (2) Next click *Add-Ins*, and then in the *Manage* box, select Excel Add-Ins; (3) Click *Go*; (4) In the *Add-Ins available* box, select the *Analysis ToolPak checkbox*, and then click OK. (If the *Analysis ToolPak* is not listed in the *Add-Ins available* box, click *Browse* to locate it. If you are prompted that the *Analysis ToolPak* is not currently installed on your computer, click *Yes* to install it.) After the

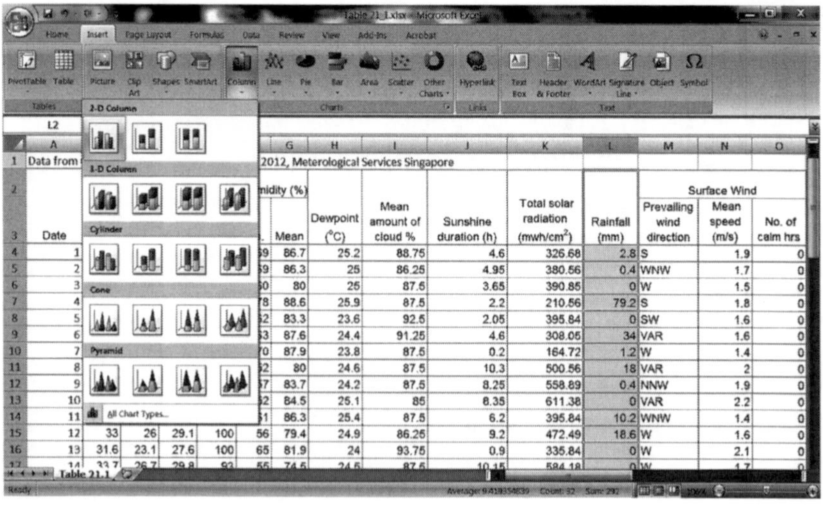

Figure 21.10 Plotting a column chart in Microsoft Office Excel.

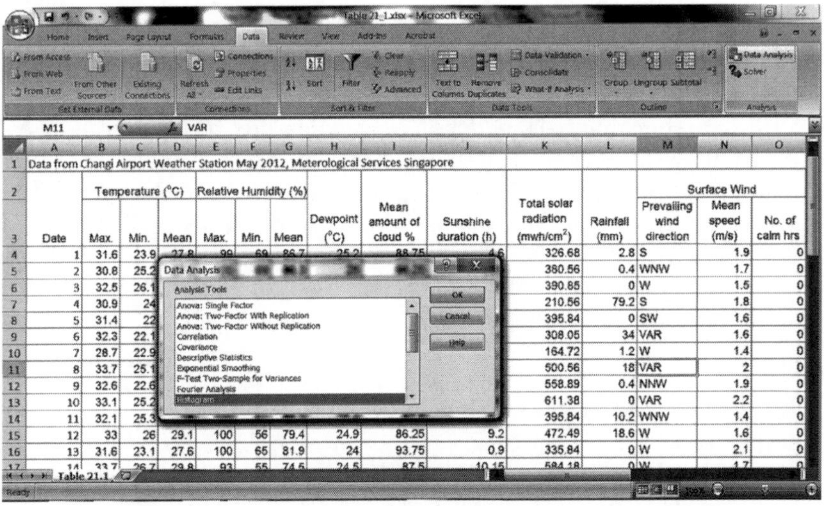

Figure 21.11 Histogram command in Data Analysis of Microsoft Office Excel.

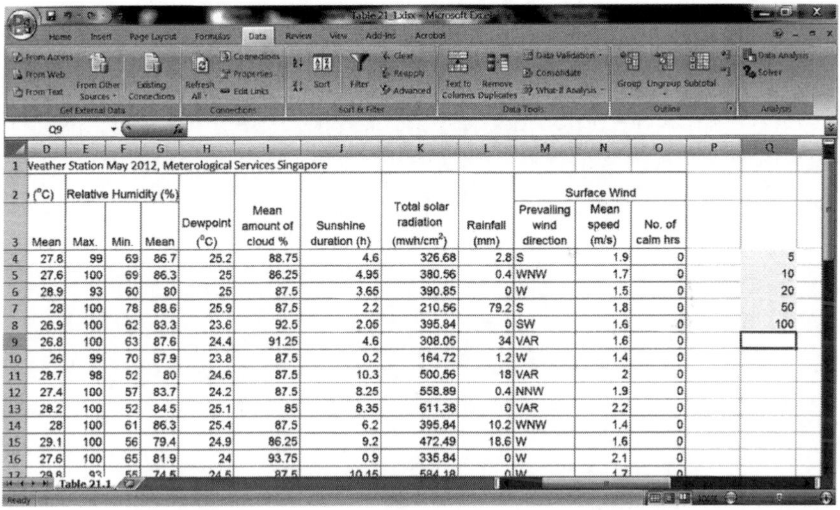

Figure 21.12 Setting up the bins values before creating a histogram in Microsoft Office Excel.

Analysis ToolPak is loaded, the *Data Analysis command* will appear in the *Analysis group* on the *Data tab*.

3. Before you can create a histogram, you will need to set the bins or intervals in a range of cells. In Table 21.3, the bins or intervals in the histogram selected are 0–5, 5–10, 10–20, 20–50 and 50–100 mm of rainfall. This is done by keying in the boundary values of the bins, 5, 10, 20, 50 and 100 into cells Q4–Q8, respectively, as shown in Figure 21.12.

4. Once the bin values are created, look at the *Data tab* and click on the *Data Analysis* command in the *Analysis group*. A dialog box should appear and histogram will be in the list. Select *histogram* and click OK. A histogram dialog box will appear as shown in Figure 21.13.

5. In the input range of the histogram dialog box mentioned in step 4 just above, select the rainfall column from cells L4 to L34. In the bin range, select cells Q4–Q8. For output option, select *New Worksheet Ply and Chart Output*. Click OK. You will see the histogram table and chart being displayed on a new worksheet as shown in Figure 21.14. The histogram can be changed to look like the histogram in Table 21.3 by changing the chart attributes, which will be explained later.

To create a bar chart:

1. Select the rainfall column from cells L3 to L34.
2. On the *Insert tab*, in *the Charts group*, click *Bar*. Select the 2D bar chart with individual horizontal bars as indicated in Figure 21.15. A bar chart of rainfalls for each day in May will be created. The bar chart will look different from that shown in Table 21.3. The bar chart in Table 21.3 requires additional formatting (see the section on *Changing attributes of a chart*).

To create a pie chart:

1. To create a pie chart, first create the data for plotting it. The pie chart is to show the proportion of sunshine hours in a day. Let's use the data for 1 May 2012. The number of sunshine hours on that day was 4.6 hours. This means that the number of hours on 1 May 2012

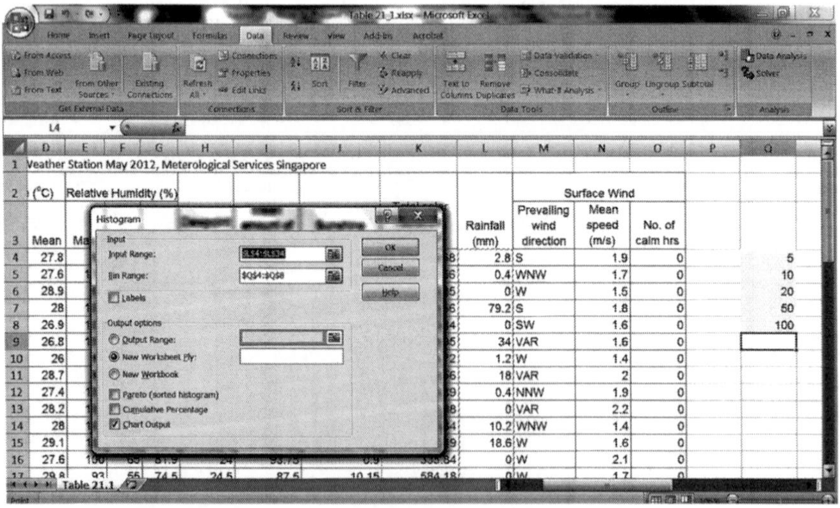

Figure 21.13 Histogram dialog box in Microsoft Office Excel.

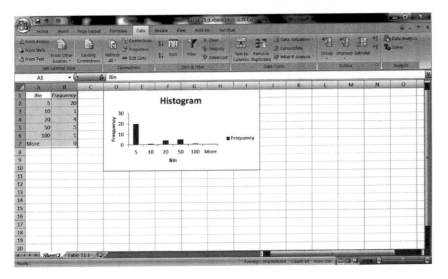

Figure 21.14 Histogram table and chart on a new worksheet.

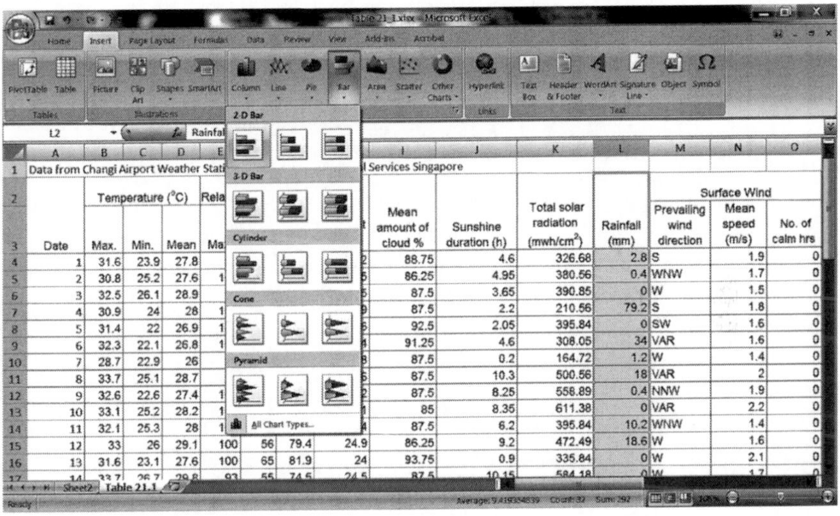

Figure 21.15 Plotting bar chart in Microsoft Office Excel.

without sunshine was (24 – 4.6) = 19.4 hours. The data are located in cells Q3–R6 in Figure 21.16.

2. Select the data from cells R4 to R5.

3. On the *Insert* tab, in the *Charts* group, click *Pie*. Select the *2D Pie* chart as indicated in Figure 21.16. You will obtain a pie chart showing the proportion of sunshine and no sunshine for 1 May

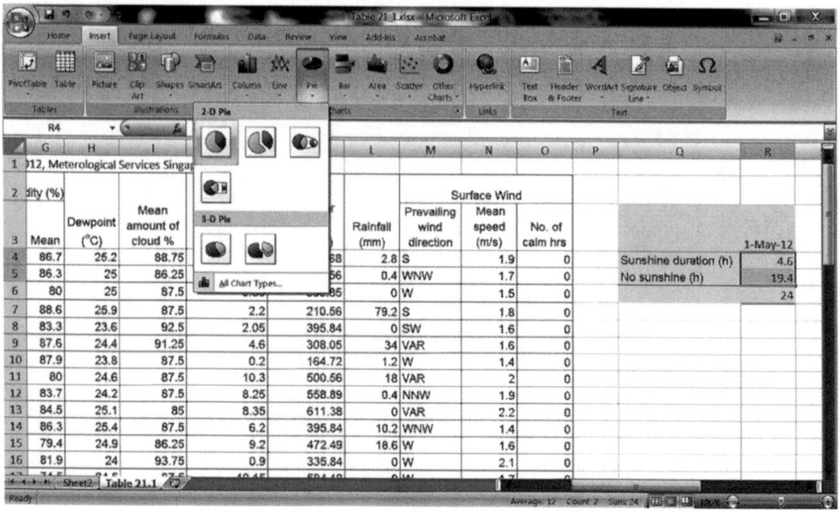

Figure 21.16 Plotting a pie chart using Microsoft Office Excel.

2012. The pie chart will need additional formatting (see section on Changing attributes of a chart) to look like the pie chart shown in Table 21.3.

Changing the attributes of a chart

The charts created in the previous section will look different from the charts shown in Table 21.3 depending on the settings of Microsoft Office Excel. The final look of the chart can be altered by changing the attributes of the chart. Before we explain how that can be done, you will need to be familiar with the anatomy of a chart as illustrated in Figure 21.17.

All components of the chart are in a finite area bound by a border. The border of the chart is not visible in Microsoft Office Excel as it shows up as a frame when selected. When selecting a chart within Microsoft Office Excel, a *Chart Tools* group will appear with three tabs: *Design*, *Layout* and *Format*. When the *Design* tab is clicked, you can change the *Chart Type* under the *Type* group, add data to the chart under the *Data* group by clicking *Select Data*, change the layout of the chart by selecting one of the icons under the *Chart Layouts* group, change the style of chart by selecting one of the icons under the *Chart Styles* group or move the chart to another location by selecting *Move Chart* under the *Location* group (see Figure 21.18). Probably the most useful command under *Chart Tools*, *Design* is the *Select Data* command which allows you to plot multiple sets of data on a single chart. Selecting data from the worksheet under the *Select Data* command is quite intuitive with the aid of a dialog box.

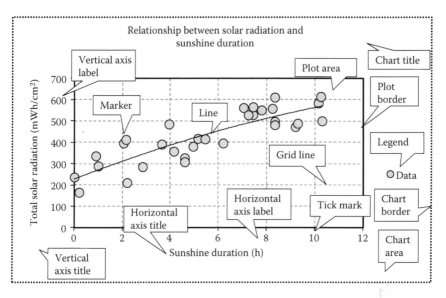

Figure 21.17 Anatomy of a chart.

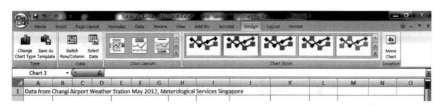

Figure 21.18 Menu items under *Chart Tools, Design* tab.

All the attributes of a chart except for its size can be changed using the *Chart Tools, Layout* tab (see Figure 21.19). The more important attributes of charts are described below:

Chart Title—can be added and positioned under the *Labels* group by clicking the *Chart Title* tab. Typically a *Chart Title* can be used in a presentation slide but it is not typical for a chart in a report or thesis. In a report or thesis, the chart is accompanied with a figure caption that is usually placed below the chart.

Axis Titles—these refer to the *Horizontal Axis Title* and the *Vertical Axis Title*. Usually a plot has only one set of horizontal and vertical axes and hence they are known as the *Primary Horizontal Axis Title* and *Primary Vertical Axis Title*. Some data from a plot may also be

Figure 21.19 Menu items under *Chart Tools, Layout* tab in Microsoft Office Excel.

plotted with a different vertical axis; the second vertical axis is then known as the *Secondary Axis*. We shall confine our discussion here to the primary axes.

Choosing the type of chart to use

Which is the most suitable chart to use? There is no absolute answer to this question. The same data can be presented using several chart types. However, the following questions can guide us when selecting a chart type to plot data:

1. Which is the chart type used for plotting similar data in other publications (e.g. academic papers and books)?
2. If the data can be presented using several chart types, which chart type is most self-explanatory?

Generally, it is best to restrict the number of series of data per chart to be between three and five. This will make the chart less crowded and more readable. If you have more than five series of data to present, consider grouping them and plotting them on more than one chart. You should also try to minimise the use of colours to differentiate one series of data from the next and rely instead on the attributes of lines, markers and shades. The true test of a good chart is how clear it is when it is reproduced on a black and white photocopier. By experimenting with different types of charts and chart layouts, you will be able to find the chart that is most suitable for your purpose.

TIPS FOR POSITIONING FIGURES, IMAGES AND CHARTS IN MICROSOFT OFFICE WORD

You may sometimes find positioning figures, images and charts to be tricky in Microsoft Office Word. The figure, image or chart may appear to go off the page when you edit part of the surrounding text. If that happens, select the figure, image and chart to bring up the *Drawing Tools Format* or *Picture Tools Format* group and select *Wrap Text* and then *Top and Bottom* as shown in Figure 21.20. This will ensure that the figure, image or chart is between two paragraphs.

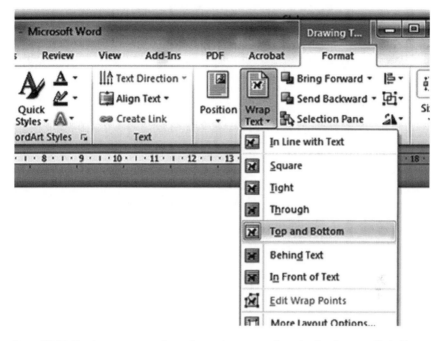

Figure 21.20 Setting text wrap for a figure, image or chart in the *Drawing Tools Format* group in Microsoft Office Word.

Another option is to create a table with a single cell without borders. Copy the figure, image or chart and paste it into the cell. The table will ensure that the figure is between two paragraphs. This is especially useful if you have multiple figures, images or charts that you would like to put into an array, for example, two figures in a column or two figures in a row. For two figures in a column, create a table with four rows and one column. Copy and paste each of the figures into the cells in rows 1 and 3. Rows 2 and 4 are for the figures' captions. Similarly for two figures in a row, create a table with two rows and two columns. Copy and paste each of the figures into the cells in row 1. Row 2 is for the figures' captions.

INTERESTING FACTS

René Descartes (1596–1650) a French philosopher, mathematician and writer, developed the Cartesian coordinates system, where paired coordinates are plotted on a plane. However, Francis Galton (1822–1911) is generally credited with the invention of the scatter plot for graphing data that we know of today.

Chapter 22

How to create a good layout

Simplicity is the ultimate sophistication.

Leonardo da Vinci (1452–1519)

Layout is the arrangement of chapters, sections, formulas, figures and tables in the report or thesis. More specifically, a layout is concerned with the placement of text, figures and tables within a page. A good layout communicates the information to the reader in a clear and effortless manner. For a book, the publisher will deal with the layout but for a report or thesis, the responsibility lies with you. As long as you follow the report or thesis format requirements of your institution, you can choose not to do anything more to enhance the layout of the thesis. However, considering the amount of time you have already spent on writing the thesis, you may find it worthwhile to follow some simple rules to make your report or thesis visually appealing to the reader. Rules for a good layout are given below.

TEN SIMPLE RULES FOR A GOOD LAYOUT

1. Finalise the layout of the report or thesis after you have completed the final revision of the report or thesis.
2. Start a new chapter on a new page.
3. Use a page fully. Do not leave blank space at the bottom of the page unless it is the last page of the chapter. You can do this by rearranging the figures and tables within the chapter to use up excess space on a page.
4. Apply the one-third rule for placing a figure or table within a page. A page is pleasing to look if the figure or table occupies one third of the page and the text, the remaining two thirds, or vice versa. The figure or table should occupy either the top of the page or the bottom of the page. Avoid embedding the figure or table in between the text or within a paragraph. The next time you read a book, observe the placement of the figures and tables on a page.

5. Place the figure or table within the page where it is being referred to in the text, or place the figure or table at the start of the following page if space does not permit.
6. Make sure that the letters and numbers in the figures and tables are not smaller than 2 mm in height so that they are legible.
7. Use no more than three colours in the same figure or table. Use colours in a figure or table that can still be read after it is photocopied using a black–white photocopier. It is better to use different line types to differentiate the plots in a figure.
8. Avoid having a table that flows over to the next page.
9. Avoid ending a page with a header (widow), or starting a new page with a line from a paragraph in the previous page (orphan).
10. Last but not least, make sure there is no spelling error in the title, chapter titles, headers, running heads, your name and your supervisor's name.

LAYOUT OF REPORT OR THESIS

Examine past reports or theses for guidelines on and requirements for these in your institution. The guidelines and requirements for the report or thesis may vary from department to department and from undergraduate level to postgraduate level. The layout for a report or thesis will generally include the following:

- The *Cover Page*, which specifies the layout (positions, font face and font size of the title of the report, university emblem, candidate's name, fulfilment statement, department, university and date). There may be a front cover page as well as an inner cover page. The front cover page usually refers to the binder cover for a soft cover page or the board for a hardcover page. The inner cover page is similar to the cover page but without the university emblem. If your final report has a hardcover, the cover page is embossed onto the board and there will also be a requirement to have the title of your project, your name and the year embossed on the spine. An example of a hardcover is shown in Figure 22.1.
- A typical *Table of Contents* page.
- Requirements for listing the examination committee, declaration of original work, abstract (or summary) and acknowledgements. Listing of the examination committee and a declaration of the original work may or may not be a requirement at your university.
- Requirements for lists of figures, tables and notations (or symbols).
- A typical page layout, includes the margins around the page, the header, the footer and the page number. An example of page layout and the various items are shown in Figure 22.2.
- The text layout, which includes the font face and the font size of the different levels of heading and the numbering system.
- Paragraph setting for first paragraph and subsequent paragraphs.

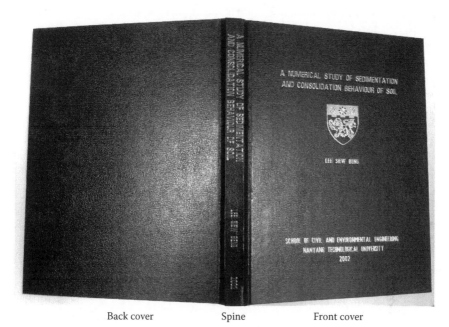

Figure 22.1 Example of a hardcover layout of a thesis.

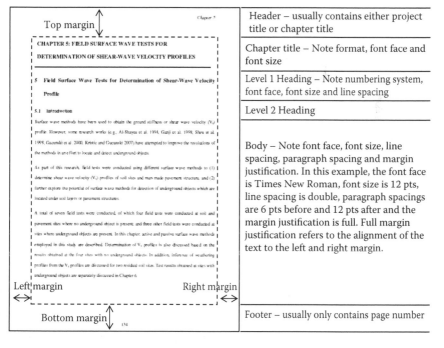

Figure 22.2 Anatomy of a page layout.

- The format of equations and the numbering system.
- The format of figures and the numbering system, placement of the captions for figures, and font and font size of the captions. The details may include the font and the minimum font size of the numbers and text within the figure.
- The format of tables and the numbering system.
- The format of the bibliography or list of references. There is a distinction between bibliography and list of references. A bibliography contains a list of publications which you refer to in your research but which you may not have cited in the text. A list of references, on the other hand, contains a list of publications that you have cited in the text. It is more common to have a list of references at the end of an academic or technical publication in engineering.

FOLLOW THE GUIDELINES

A report or thesis is a formal document and is written in a particular format imposed by the institution. Check with your institution for the format requirements of your thesis. Your university may provide a template file with the format requirements built in for your use. It will save you a lot of time if you use this template file when you write as the template file will ensure that the format of each chapter is consistent. The template file will also enable you to generate a *List of Tables, List of Figures and Table of Contents* effortlessly. If the template file is not used correctly, you will encounter problems later on, which may lead to time wasted; in the worst case, your files may get corrupted.

CREATE A TEMPLATE FILE

If a template file is not provided by your university, you may want to create your own template file before you start writing. Creating a template file for a whole report or thesis is more demanding than creating a template file for a chapter. It is usually sufficient to create a template file for a chapter as you will be using it in all your chapters. You can create a template file from an existing template, or from a new or existing Microsoft Office document. We will use the simplest method, which is to create a template file for a chapter from a new document in Microsoft Office Word. The default template file when you start a new document in Microsoft Office Word is the Normal template and is stored as normal.dot for Word 97–2003 or normal .dotx for Word 2007 and 2010. To create the template file:

1. Open a new document in Microsoft Office Word.
2. Set the layout of the page by selecting the *Page Layout* tab (Figure 22.3). Select *Margins* and select the option according to your institution's

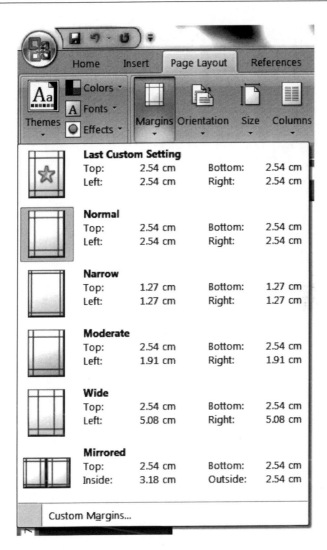

Figure 22.3 Setting margins in Microsoft Office Word.

guidelines. Select *Size* and choose the correct paper size. The usual paper size in the US is *Letter* size whereas in most other parts of the world, it is *A4* size. It is assumed that the default *Orientation* is set at *Portrait* and *Columns* is set as *One*. If not, you may want to select the correct setting by selecting *Orientation* and then *Columns*.

3. Set the format of the header and footer of a page if there are special requirements. Select the *Insert* tab and then click on *Header* to select any of the *Built-In* headers meet your requirements (Figure 22.4). If there are none, you can select the *Built-In* header that is closest to

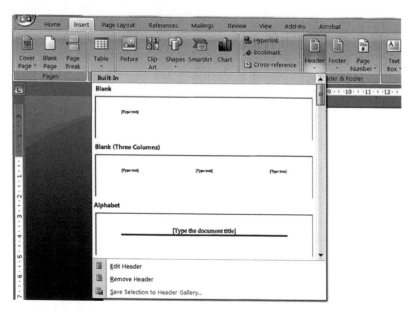

Figure 22.4 Inserting a header in Microsoft Office Word.

your requirements and *Edit* it afterwards. You can format the *Footer* in the same manner as for the *Header*.

4. Set the page number by selecting the *Insert* tab and then click on *Page Number*. It is in the same group as *Header & Footer*. Select the position of the page number as shown in Figure 22.5.

5. Format the title, headings and captions in a chapter. The title, headings and captions in a chapter are formatted in the *Styles* group. *Styles* indicate how the text should appear in your Word document. Under the *Home* tab, you will find a *Styles* group. If you click on the bottom right of the *Styles* group (circle 1), you will get the *Styles* pane as shown in Figure 22.6. This gives a vertical listing of the many styles you see in the *Styles* group, for example *Normal, No Spacing, Heading 1, Title* and *Subtitle*.

The styles that you will need in a chapter are chapter title, first paragraph (of a section), normal paragraph, list paragraph, headings (levels 1, 2, 3 etc.), figure caption, table caption and equation. Setting the style for each of them is similar. We shall explain how to do it for a normal paragraph. In the *Styles* pane, click the *Normal* style then click the *New Styles* button (circle 2) on the bottom right of the *Styles* pane as indicated in Figure 22.6 to bring up the *Create New Style from Formatting* dialog box. You can then create a new style. You may want to rename the style in

Figure 22.5 Inserting page numbers in Microsoft Office Word.

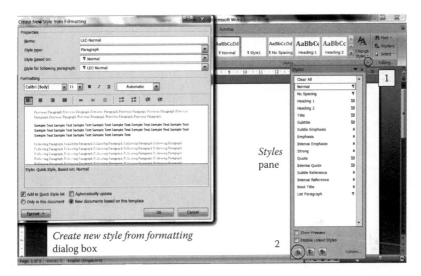

Figure 22.6 Creating a new style from the *Normal* style in Microsoft Office Word.

Name using your initials as a prefix (say LEC-Normal) so that you will be sure that it belongs to you and is not from some other template. This style is automatically based on the *Normal* style as you had selected the *Normal* style to create the new style. You can select the default style following a paragraph you typed with the LEC-Normal style in the *Style for following paragraph* dropdown box. Select the radio button *New documents based on this template*. Change the attributes of the *Styles* by selecting the *Format* button (see Figure 22.7). You can change the font, paragraph, tabs and numbering (for a list paragraph) as shown in Figure 22.7.

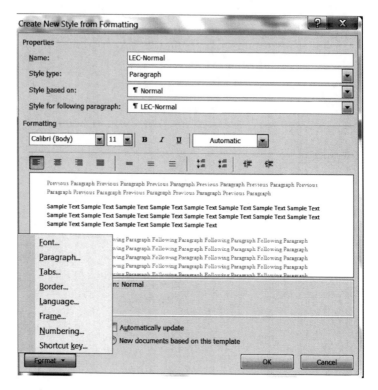

Figure 22.7 Modifying the style in Microsoft Office Word.

6. Save the file as a Word Template (.dot for Word 97–2003 or .dotx for Word 2007, 2010).

7. When you want to start writing a new chapter, open the template file and save the file as a new Word document first before you start typing. Select the appropriate style before you type. Say you want to type the chapter title; select the chapter title style and type the title. The default style should always be the normal paragraph style. If you want to type a new section, select *Heading 1* for the main section heading; select *Heading 2* if you want to type a new subsection and so forth. As far as possible, try to avoid formatting font and paragraph without using one of the *Styles*.

Using a template file and using *Styles* in your typing will enable you to generate the *Table of Contents* page effortlessly. To do this go to the location you want to place the *Table of Contents*, select the *References* tab and then select *Table of Contents* to select the format style of the table of

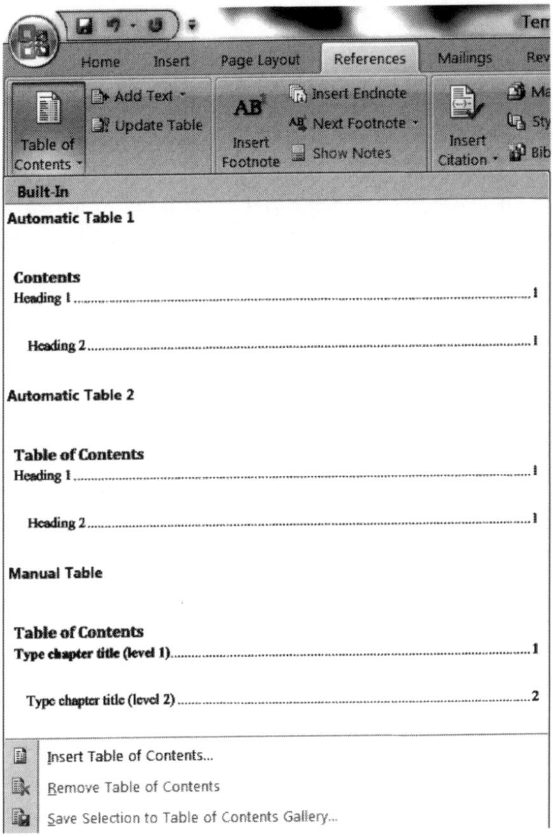

Figure 22.8 Creating a Table of Contents in Microsoft Office Word.

contents that meets your requirements (Figure 22.8). When you are more comfortable with using a template file, you may want to explore modifying the *Styles* in your template file to achieve a more visually pleasing layout of your report or thesis. There are many articles on modifying and creating a template file on the Internet, for example https://support.office.com/en-SG/Article/Word-2010-Create-a-new-template-cb17846d-ecec-49d4-82ea-a6f5e3e8b9ae.

If you accidentally save your template file as the Word default template file (Normal.dot in Word 97–2003 or Normal.dotx and Normal.dotm in Word 2007, 2010) and want to restore the default template file, exit Word, locate the Normal.dot or Normal.dotx or Normal.dotm file and delete it. When you start Word again, the original default template file will be re-created.

INTERESTING FACTS

The earliest book with a layout is known as an incunabulum (Meggs and Purvis, 2012). It is handwritten, usually in a beautifully designed layout. When the printing press was invented by Johann Gutenberg in 1450, the layout imitated those of the incunabula. The Gutenberg Bible of 1455 is an incunabulum. The most well-known incunabulum is the Nuremberg Chronicle, which gives an illustrated history of the world divided into seven ages. It was written in Latin by Hartmann Schedel in 1493.

REFERENCE

Meggs, P. B., & Purvis, A. W. (2012). *Meggs' History of Graphic Design* (5th ed.). Hoboken, NJ: John Wiley & Sons.

Chapter 23

How to prepare for an oral presentation

Give me six hours to chop down a tree and I will spend the first four sharpening the axe.

Abraham Lincoln (1809–1865)

A presentation provides an opportunity for you to communicate your research findings orally. Research papers and reports are usually presented in seminars, conferences or poster presentations. Such sharing of information provides an opportunity for researchers to receive feedback at crucial junctures in the project. Many people associate presentation with a collection of Microsoft Office PowerPoint slides. This has spawned a number of phrases such as 'Death by PowerPoint', 'PowerPoint hell' and 'PowerPoint poisoning' to describe presentations that used slides poorly. Similar phrases also apply if you are not using other presentation software effectively. The purpose of a presentation is for you to relay a message to the audience. The difference between a good and a bad presentation is determined by the effectiveness of the presenter in getting the message across. Presentation materials are there to help the presenter get his message across and not be the message itself. The relationship between the presenter, presentation material and audience is shown in Figure 23.1.

THE MESSAGE

The message in your presentation is your research findings. All the materials that you need for your presentation can be extracted from your report or thesis. You should not present any material that is not part of your report or thesis. An effective presentation requires careful selection of the materials from the report or thesis. Common mistakes in selecting materials are (1) including too much information, especially including slide after slide of equations that are difficult for the audience to process and understand and (2) including too much of other people's work (literature review) and not enough of your own work.

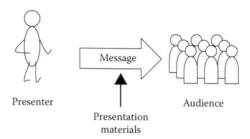

Figure 23.1 Relationship among presenter, presentation materials and audience.

THE AUDIENCE

The audience attending a research presentation is less varied than those attending a general topic presentation. You can expect some of them to be experts in your field, or on your topic. The reason for you making the presentation is to present the key findings and conclusions from your research to the audience. The audience decides the success of your presentation. If you are presenting to defend your report or thesis, the audience 'examines' your work. It is thus important to select presentation materials that meet the expectations of your audience. Avoid presenting materials that are deemed common knowledge in your field.

PRESENTATION TIME

It is important to keep within the allocated time. It is better to use less of the time allotted than to exceed it in a presentation. As a rule of thumb, you should have about one slide per minute of your presentation time. This rule forces you to be more selective of the presentation materials to include as well as avoid having too many slides.

PREPARING FOR AN ORAL PRESENTATION

For an effective presentation, prepare thoroughly using the following four steps:

1. *Plan* the materials that you are going to include in your presentation. You will know what materials that you should include by asking yourself these four questions: (1) Why did I do this work (emphasis on 'this')? (2) How did I do it (what tools, techniques, approaches were used)? (3) What were my findings (results)? (4) What have I concluded from it (what does it all mean)?

2. *Organise.* Arrange the materials in the order that you shall be presenting. Think about the organisation as building a storyboard for your presentation.
3. *Support with visuals.* A figure or image is more attractive than a slide of text, a single equation is better than a slide of equations and less text (point forms) is more effective than more text. The current trend in presentation is to reduce the number of slides to the minimum and keep their design simple. This approach is called 'Presentation Zen' (Reynolds, 2011).
4. *Practise.* All great presenters such as Steve Jobs spend an enormous amount of time practising their presentation. With enough practise, you can present without the need to memorise the content.

PRESENTATION FORMAT

Use the report or thesis framework for your presentation:

- Introduction slides
- Materials and methods slides
- Results and discussion slides
- Conclusions and future research slides
- References cited slides

Introduction slides

- State the problem, the objectives of the study, and brief background information.
- Include the justification and relevance of your study.
- Try to answer the following questions: Why was the study done? What is the existing state of knowledge of this topic? What are the specific objectives?
- Clearly state the research question that you sought to answer.

Materials and Methods slides

- Include a brief description of the procedure you used.

This will include data collection techniques, subjects studied, location of the study and the methods used to record, summarise and analyse the data.

Results and Discussion slides

- Present your most important results.
- Use equations, tables and figures as appropriate.

- Include an interpretation and evaluation of the results.
- Compare results with those from other studies (if appropriate).

Conclusions slides

- Draw conclusions based on your findings.
- Speculate on the broader meanings of the conclusions drawn.
- Identify sources of error and any inadequacies of your study.
- Address any future study that your research suggests (recommendations).

References Cited slides

- List all the references cited in the text—if you did not cite any references, you do not need a *References Cited* slide.
- Cite references in text by author and date.
- All references should be listed in full, alphabetically by first author, in the *References Cited* slides.

PRESENTATION SLIDES

Slides are designed to be visible and to be read by the audience. They should be clear, concise and succinct. Here are some simple rules for slide design:

1. Use large fonts (at least 24 points). The rule of thumb is that the person at the back of the room should be able to read the slide.
2. Choose san-serifs fonts (e.g. Arial, Calibri, Helvetica) over serifs fonts (e.g. Times New Roman, Century, Lucida) unless otherwise stipulated by your institution as research has shown that it is harder to read serifs fonts.
3. Use boldface or underline for emphasis and minimise the use of italics, which is harder to read.
4. Use a dark background with light fonts and graphics, or use a light background with dark fonts and graphics.
5. Do not write complete sentences. Reduce and simplify text by using a bullet list.
6. Include no more than four to five bullet points on a single slide.
7. Make all the bullet points on a single slide relevant to one single specific point.
8. Use a heading for each slide.
9. Avoid long complex sentences—break down into relevant subcomponents, each with a separate bulleted entry.
10. Use active voice.
11. Avoid using all capitals in the text as this comes across as being rude.

12. When using colours on a slide, use no more than three colours. Avoid certain colour combinations such as red–green (difficult to decipher for people with red–green colour blindness), yellow–white or blue–black (difficult to see due to low contrast) and red–blue (can create illusion and causes eye strain and fatigue).
13. When presenting figures and charts, make sure captions and labels are visible.
14. Minimise the use of animations or transitions as they can be distracting. However, animation can be useful for emphasising points and animating flowcharts.
15. Avoid putting too much information on one slide. Use a presentation script, if necessary.

A common mistake made by students, especially those whose first language is not English, is to assume that the slides will be sufficient cues for the oral presentation. However, the anxiety and emotions experienced during the actual presentation may result in a disconnect between the brain and tongue, leading to poor choice of words and disjointed sentences. Preparing a script should be part of the preparation for the presentation. Many public speakers use a prepared script. If you do not see them using a script, it is because they have committed the script to memory after countless rehearsals or they were using a teleprompter. Therefore, prepare a script even if the presentation is only for 10 minutes. Write the script using spoken English and simplify the sentences to make your presentation as effective as possible. Your presentation should sound like a conversation with the audience.

Once you have prepared the script, rehearse, rehearse and rehearse! If you have rehearsed sufficiently, you should be able to do without the script during the presentation. If you need to use the script during the presentation, take only quick glances at it while maintaining sufficient eye contact with the audience.

INTERESTING FACTS

- According to Weissman (2009), in an oral presentation how you present yourself (body language) is the most important (55%) followed by how you deliver the presentation (38%) and lastly the words you used in the presentation (7%).
- The 10/20/30 Powerpoint rule—This rule was devised by Guy Kawasaki, a venture capitalist, author and speaker. The rule is that no presentation should have more than 10 slides, uses less than 20 minutes and font size of at least 30 points. This rule is useful for any presentation lasting less than 20 minutes.

REFERENCES

Reynolds, G. (2011). *Presentation Zen: Simple Ideas on Presentation Design and Delivery* (2nd ed.). Berkeley, CA: New Riders.

Weissman, J. (2009). *The Power Presenter: Techniques, Style and Strategy from America's Top Speaking Coach*. Hoboken, NJ: John Wiley & Sons.

Chapter 24

Do's and don'ts of oral presentations

To me, presentations are the most powerful device. You can't really name a movement that didn't start with the spoken word.

Nancy Duarte, author, *Slide:ology: The Art and Science of Creating Great Presentations, 2005*

After you have written and submitted your report or thesis you are expected to present it orally to your examiners or to a review board. The oral exam allows you to demonstrate your understanding of your research, the related literature and the topic as a whole. It gives your examiners the chance to question you on any areas in your work that are not clear to them, and also evaluate whether the work is your own. The oral exam can thus be stressful especially if this is the first time that you will be presenting your work orally to an audience who are experts in your area of research. It can be even more stressful if English is a second or foreign language for you.

The objective of this chapter is to provide you with practical advice on how to present your work confidently and effectively. Read *Chapter 23 Preparing for a project presentation* for advice on how to prepare the contents of your presentation. This chapter covers the following areas in delivering an oral presentation successfully:

- Managing anxiety
- Delivering your presentation
- Using presentation tools
- Managing your time
- Answering questions

MANAGING ANXIETY

Do:

- *Learn to accept some anxiety.* Even the most experienced speakers feel a little nervous before a presentation; in fact, it is believed that a little anxiety actually makes you a better speaker. Learn to accept that you will always be a little anxious about speaking in public, and that it is normal and common to feel this way.
- *Visualise confidence.* Visualise yourself confidently delivering your speech. Imagine giving the presentation feeling free of anxiety and engaging the audience. Although this may appear unrealistic for you now, visualisation is recognised as a powerful tool for changing the way that we feel. Elite athletes use this strategy to improve performance in competitions and so can speakers to improve their delivery.
- *Practise.* Even great speakers practise their speeches beforehand. Do practise your presentation at least a couple of times before you present it. Practise in an environment that is as similar as possible to where you will be giving your presentation. Practise using your visuals/PowerPoint slides. Practise out loud with a recording device or video camera and then watch yourself to see how you can improve. Or, practise in front of friends or family members and ask for feedback.
- *Memorise the first minute.* Start with your brain on autopilot by knowing the first minute of your presentation by heart and knowing the rest of your presentation very well.
- *Write out the script of your presentation.* Using a complete script will help if you are extremely nervous about presenting. Practise reading it in a natural voice (you do not want to sound like a textbook). During your presentation, refer to your script discreetly when you are speaking. Hold the script at waist level and take only brief glances at it.
- *Visit the venue.* If you have access to the room where you will be speaking, take the time to visit in advance and get used to the layout of the room in which you will be presenting. Check out the audiovisual equipment and make arrangements for any equipment that you will need but is not available. Practise standing where you will deliver your speech. Familiarising yourself with your surroundings will take away the fear of the unknown and help to reduce your anxiety.
- *Concentrate on your key points.* When you focus on the task at hand, anxiety is less likely to get out of control. Concentrate on the main points of your presentation and make it your goal to deliver those key points to your audience.
- *Realise the audience is on your side.* Think about a time when you were a member of an audience and the speaker was noticeably nervous. Did you think negatively of that person? More likely, you felt sympathetic and attempted to reassure the speaker by smiling or

nodding. Remember that in most situations, the audience wants you to succeed.

- *Breathe.* Relax and breathe normally. Do not be afraid to pause and take a deep breath or two before you begin (or during the presentation, if necessary).
- *Manage symptoms of nervousness.* During your presentation, deal with symptoms as they occur:
  Dry mouth: sip a little water
  Trembling hands: clasp them lightly together
  Shaking knees: shift your weight and flex your knees
  Quivering voice: pause, take a deep breath or two, and smile to relax

Don't:

- *Rely completely on a written script.* If you are very nervous, it helps to bring along a script of your presentation as a form of security, in case you forget what to say. However, beware that when people are nervous, they tend to just read the script without looking at the audience. This makes you look unprepared as well as unprofessional. Unless you are extremely nervous and need to rely on a written script, put what you need to remember on your slides as brief points. Then use these as prompts to remind you of what you have planned to say. Remember, however, to glance only briefly at your prompts on the screen.
- *Limit practise to just before the oral examination.* Take every opportunity to speak in front of groups and to give presentations throughout your programme of studies. It is only by frequent practise that you will overcome your nervousness about speaking in public.

DELIVERING YOUR PRESENTATION

A presentation comprises both content and delivery. Content has to do with the *what*, that is the information being conveyed. Delivery has to do with the *how*, that is the manner in which the content is conveyed. Delivery has three components: *verbal, vocal, visual*; these are often referred to as the 3 *V's*. The *verbal* component relates to the use of words, the *vocal* component to how we use our voice, and the *visual* component to the different aspects of body language.

Verbal component: Use of words

Do:

- *Open and close the presentation in a professional manner.* Treat the beginning and the end of your presentation seriously. The beginning is when you establish rapport with the audience and when you have its full attention. The end of your presentation is just as important as it leaves a lasting impression.

Open your presentation by greeting your audience, introducing yourself, showing your title slide, and looking directly at the audience. Say: *Good morning/afternoon. I am (your name). I'm going to present my project/research on ...* Then move smoothly on to the next slide.

Close your presentation by projecting your *Conclusions* slide. Run through the points on the slide very briefly. Then nod your head and thank your audience.

- *Use spoken English style, not written.* Speak in the style of a relaxed but serious conversation. You should give the impression that you are speaking to the audience, rather than sounding as though you are reading your report or thesis.
- *Speak in the active voice.* Unless you have a reason to use the passive voice, speak in the active voice as this is more natural in spoken English. For example, instead of saying *A correlation was found between ...* (passive voice), say *We found a correlation between ...* (active voice).
- Use the correct technical terms.
- *Use simple and concise words.* Avoid unnecessarily unfamiliar or complicated words. For example, say *read* rather than *peruse*; *help* rather than *facilitate*; *explain* rather than *elucidate*.
- *Keep sentences short and concise.* Avoid long and convoluted sentences. Pause at the end of sentences.
- *Give verbal signals.* Use expressions to highlight, emphasise important information or indicate relationships between ideas such as: *This was an interesting result as ...; This was unexpected because ...; This was, in fact, higher than ...* and so on. Verbal signals also include transitions which indicate the beginning of a new point, examples of which are *Firstly, ...; Following from ...; Moving on ...*

Don't:

- *Use slang or informal language.* In a formal presentation, avoid words such as *like, OK, right, things, stuff, hassle* and so on.
- *Memorise your paper* and deliver it verbatim. You may want to memorise the introduction and conclusion so that you start and end on a strong note, but your goal is to sound natural. Reading sentences does not sound spontaneous and engaging.
- *Read your PowerPoint slides.* Reading your slides to the audience is not only redundant but makes your presentation extremely boring. Elaborate and discuss rather than read what is on the slide. You need to make eye contact with your audience and you cannot do that if you are reading the slides.

- *Read sub-headings.* A common mistake pointed out by Silyn-Roberts (2013, p. 243) is to read sub-headings on the slides, for example:
  Research objectives. Research objectives were ...
  Sampling methods. Three sampling methods were ...
  Avoid reading sub-headings as it gives the impression that you are reading from a script or that you are insufficiently prepared.

Vocal component: Use of our voice

Do:

- *Rehearse your presentation using an audio or video recorder.* Listen to the recording and note the tone, pitch and speed of your voice. Work on sounding natural and relaxed.
- *Pronounce your words correctly and clearly.* If you are unsure how a word is pronounced, check the pronunciation before the presentation. Ask a fluent speaker of English for help or use an online dictionary, for example www.dictionary.com or www.howjsay.com. Mispronouncing key words in your field may give the audience a negative impression of you and affect your credibility.
- *Articulate your words clearly.* Do not drop syllables (saying *techlogy* instead of *technology*), swallow your words or garble them.
- *Minimise using fillers.* Fillers are vocal distractors *(um, ah, er)* or verbal distractors which are repetition of the same word *(ok, right, actually, basically, you know etc.)*. There is nothing wrong with remaining silent while you think of what to say.
- *Vary the tone of your voice to keep your audience's attention.* It is difficult to pay attention when the presenter is speaking in a monotonous voice. Make your voice interesting by varying volume, rate of speech or pitch.
- *Express enthusiasm for your subject.* Enthusiasm is an essential element in any successful speaker's delivery. To speak enthusiastically, you need to *feel* that you have something important to say and that what you have to say is significant for your audience. Enthusiasm also requires an investment of *energy*. You need to invest more energy than when you are engaging in daily conversation.
  Note: Watch a video of Steve Jobs giving a presentation for an example of an enthusiastic delivery: www.youtube.com/watch?hl=en-GB&gl=SG&v=2-ntLGOyHw4.
- *Emphasise key words in your presentation to heighten the impact of your message.* Before the presentation, identify 10–20 words in your presentation to emphasise. Techniques for emphasis include saying the word by changing the volume (speaking the word louder or softer), by raising the pitch (saying the word in a higher pitch) or by saying the word a little longer (stretching out the word, e.g. *en-vi-ron-ment*).

- *Pause at the end of sentences.* Pauses give your audience time to absorb information. Pause before important information to maximise its impact.

Don't:

- *Speak too quickly.* Too rapid speed makes it difficult for your audience to process a lot of new information.
- Speak in a monotone or mumble.

Visual component: Body language

Do:

- *Rehearse your presentation using a video recorder.* Watch the video recording and note body language and facial expression. Work on appearing confident and relaxed.
- *Dress appropriately for the presentation.* Your personal appearance affects your credibility. Casual clothing such as jeans and T-shirts are not appropriate for a formal presentation.
- *Make eye contact with members of the audience.* Sustain eye contact with different members of the audience for 4–5 seconds to appear confident and interested.

Don't:

- *Slouch or lean to one side.* Stand straight and distribute your weight evenly on both legs.
- *Sway or rock in place.*
- *Fidget.* Examples of fidgeting are fiddling with a pen, clicking a pen or tapping a foot.
- *Pace back and forth aimlessly.* Moving about excessively is distracting to the audience. Make your movements purposeful.

USING PRESENTATION TOOLS

- *Note cards.* Write in large, boldface letters if you are using note cards or regular paper for your notes so that you can read your notes quickly and easily. To engage your audience, you should look up from your note cards or paper several times during the presentation.
- *Interaction with the screen.* Be aware of the audience and do not block their view. Do not look at the screen too much. The ratio should be 30:70—look at the screen 30% of the time and at the audience 70% of the time. Point out important features in the figures, graphs

and other visual aids to the audience. Your audience should not be expected to navigate their own way through a complex diagram.

* *Using a laser pointer.* If you are using a laser pointer, make sure you do not continuously and aimlessly circle it on the screen. Locate the word or a particular point in a visual you want to highlight and keep the pointer firmly in the right place for 1 or 2 seconds.

MANAGING YOUR TIME

People often run out of time when giving a presentation and then rush to complete it. The following are the most common reasons for running out of time and tips for avoiding this as suggested by Silyn-Roberts (2013).

Don't:

1. *Spend too long on the introductory material.* Your audience is primarily interested in hearing about your results and conclusions. So do not go into great detail about other people's work. Give just enough background information for the audience to understand the context and motivation of your research.
2. *Explain some slides more than you planned to.* It is very easy for unplanned material to come into your mind in the stress of the moment. Resist this by planning the points you need to make for each slide, and forcing yourself to stick to them.
3. *Practise your presentation too fast.* When practising or rehearsing your presentation, remember to speak it out loud. If you merely read your presentation or whisper it, you go faster than you would when you speak it.

PROCEDURE FOR FINISHING QUICKLY

If you do run out of time and need to finish quickly, Silyn-Roberts (2013) advised that you follow the steps below to help you deliver the main points of your presentation to your audience without rushing and in a professional manner (Silyn-Roberts, 2013, p. 247):

Step 1: When you realise that time is running out, do not panic. Stay calm and avoid apologising or saying anything to draw attention to the situation. Finish the sentence you are saying smoothly and go directly to your *Conclusions* slide.

Step 2: When the *Conclusions* slide is displayed, make a brief comment such as, 'I'm afraid I don't have the time to go through all of my material. However, the conclusions from this project/research are ...' Then run through the conclusions briefly.

If the time left is really very short, show the *Conclusions* slide for the audience to read and then close the presentation by thanking the audience.

TIP: HOW TO FIND YOUR CONCLUSIONS SLIDE INSTANTLY

If you are using PowerPoint, note down the number of the *Conclusions* slide before the presentation. When you need to show the slide, just type the number and hit the *Enter* key.

ANSWERING QUESTIONS

Many students are nervous at the thought of answering questions at the end of a presentation because they fear not knowing the answer to the questions. Students for whom English is a foreign language may also fear not understanding the questions they are asked.

The following suggestions given by Zwickel and Pfeiffer (2006) to help you respond to questions effectively:

1. *Work out possible questions beforehand.* Discuss with your supervisor or someone familiar with your work the questions that are likely to be asked. Anticipate questions on weak or problematical areas and work out how to answer them.
2. *Make sure you understand the question correctly.* Do not answer a question until you are sure you have understood it. Ask for clarification if necessary. You could say, for example: *I'm sorry, I'm not sure I understand that. Could you repeat it, please?*
3. *Repeat the question if you think the audience may not have heard it.* Say: The question was *What is the most likely cause of ...?*, and then answer it.
4. *Prepare a supplementary set of slides.* These slides can be more detailed than those in your main presentation. Show them if you are asked a question that requires more detail than provided in the main presentation.

Don't:

1. *Pretend if you do not know the answer to a question.* An expert audience can easily see when someone is unable to answer a question and attempts to cover this up. It is better to be honest by either saying that you do not know (in a positive voice), for example, *I'm afraid I don't know the answer to that question*, or offer to find out the answer, for example, *I don't have the answer to that question at the moment, but*

I'll find out for you by.... Besides indicating honesty, such responses also show openness to communicating about your work and your confidence in it.

2. *Interrupt people when they have not completed asking their questions.* Besides being rude, the question may be misunderstood. It is a good idea to wait 1 or 2 seconds after the question has been asked before answering it. This allows you to process the question properly and time to frame an appropriate response.

3. *Be negative about your work.* Do not be afraid to mention briefly and objectively difficulties you may have encountered in your research. It is a sign of honesty, not of weakness. The audience will be able to relate to it and someone may be able to advise you. However, do not be overly critical or self-pitying.

CHECKLIST FOR A PRESENTATION

Delivering your presentation

- ☐ Is the volume of your voice sufficient?
- ☐ Is your rate of speech appropriate?
- ☐ Do you sound enthusiastic about your work?
- ☐ Are your words correctly and clearly pronounced?
- ☐ Do you make sufficient eye contact with the audience?
- ☐ Are you standing straight and confidently?
- ☐ Are your gestures appropriate and natural?

Answering questions

- ☐ Have you worked out beforehand the possible questions?
- ☐ Do you know how to deal with questions you do not understand?
- ☐ Do you know what to say when you do not know the answer to a question?

INTERESTING FACTS

The greatest fear

According to U.S. surveys, the fear of public speaking ranks higher in most people's minds than fear of spiders, snakes, illness, heights and even death itself. Even professionals suffer at times from speech anxiety. Sir Lawrence Olivier, a famous British actor, suffered from extreme stage fright.

The longest, shortest and the speech that killed

William Henry Harrison, the ninth U.S. president, is credited for giving the longest speech at an inaugural ceremony. The speech comprising 8,445 words was given in cold and wet weather and resulted in him catching pneumonia, which killed him a month later. George Washington, the first U.S. president, on the other hand, gave an inaugural speech that was just 135 words long. The record for the longest speech was set in 1828 by the House of Commons (the lower house of Parliament in the United Kingdom), and still stands. It lasted 6 hours.

Commercial impact

Three out of every four people, amounting to 75% of the population, suffer from fear of public speaking. Not surprisingly, self-help books offering advice on speaking in public are the best-selling self-help books.

(Extracted from Public Speaking Success, n.d.)

REFERENCES

Public Speaking Success. (n.d.). *Fun Facts and Statistics on Public Speaking.* Retrieved from: http://publicspeakingsuccess.net/fun-facts-and-statistics-on-public-speaking/.

Silyn-Roberts, H. (2013). *Writing for Science and Engineering: Papers, Presentations and Reports* (2nd ed.). Waltham, MA: Elsevier.

Zwickel, S. B., & Pfeiffer, W. S. (2006). *Pocket Guide to Technical Presentations and Professional Speaking.* Upper Saddle River, NJ: Pearson Prentice Hall.

Appendix: Common editing symbols

Symbol	Meaning	Example
_e_	Delete	Write your ~~final~~ report.
		Write your report.
‿	Close up	Here in is a diagram.
		Herein is a diagram.
_e_	Delete and close up	Revsise the draft.
		Revise the draft.
∧	Insert	not We did ∧ study it.
		We did not study it.
#	Space	The#study was finished.
		The study was finished.
∿	Transpose	Pour in the mix liquid.
		Pour in the liquid mix.
/ or lc	Lower case	The Technician agreed.
		The technician agreed.
≡	Capitalise	See Table 4.2.
		See TABLE 4.2.

	Capitalise first letter and lower case remainder	The E̷ARTH is polluted.
		The Earth is polluted.
stet	Ignore correction made	Close the blue door. (stet)
		Close the blue door.
¶	New paragraph	The study ended. ¶A new phase has started.
		The study ended. A new phase has started.
no ¶	Remove paragraph break	This process stopped it. no ¶ It progressed to the next stage.
		This process stopped it. It progressed to the next stage.
	Move to a new position	The participants attended who were men
		The participants who were men attended.
	Flush left	Refer to Figure 10.
		Refer to Figure 10.
	Flush right	Refer to Figure 11.
		Refer to Figure 11.
	Centre	Figure 8.2
		Figure 8.2

∧	Superscript	36∧5
		36^5
∨	Subscript	36∨5
		36_5
⊙	Period	List the steps⊙ Next, select five of the steps.
		List the steps. Next, select five of the steps.
∨'	Apostrophe or single quote	The job∨' scope is shown.
		The job's scope is shown.
∧⌣	Semicolon	He left∧⌣Yet, he chose to return a few days later.
		He left; yet, he chose to return a few days later.
∧⌣	Colon	There are three parts∧lock, stock and barrel.
		There are three parts: lock, stock and barrel.
∧⌣	Comma	Choose one∧two or three steps.
		Choose one, two or three steps.
∧_	Hyphen	A by∧product is created.
		A by-product is created.
∧pg∧	Page number	∧pg∧
		15

◯ sp	Spell out	The ⊙phone⊙ sp malfunctioned.
		The telephone malfunctioned.
◯ ab	Abbreviate	The line is ⊙fifteen⊙ ab ⊙centimetre long⊙. ab
		The line is 15 cm long.
‖	Align	Participants Experimental variables ‖ Materials
		Participants Experimental variables Materials
	Underlined	See <u>Table 2</u>.
		See <u>Table 2</u>.
	Join to previous line	They revised it and submitted the draft.
		They revised it and submitted the draft.
ital	Set in italics	Data analysis ital
		Data analysis
bf	Set in bold	Chapter 1 bf
		Chapter 1
wf	Wrong font	The **steps are clear** wf
		The steps are clear.

Index